Fuels and Petrochemicals Division

Presentations at the 2008 AIChE Annual Meeting

Philadelphia, Pennsylvania
16 - 21 November 2008

ISBN: 978-1-61567-229-5

Printed from e-media with permission by:

Curran Associates, Inc.
57 Morehouse Lane
Red Hook, NY 12571

CURRAN ASSOCIATES INC.
proceedings.com

Some format issues inherent in the e-media version may also appear in this print version.

Printed by Curran Associates, Inc. (2009)

AIChE
3 Park Avenue
New York, NY 10016-5991

Phone: (203) 702-7660
Fax: (203) 775-5177

www.aiche.org

Additional copies of this publication are available from:

Curran Associates, Inc.
57 Morehouse Lane
Red Hook, NY 12571 USA
Phone: 845-758-0400
Fax: 845-758-2634
Email: curran@proceedings.com
Web: www.proceedings.com

TABLE OF CONTENTS

Growing Lipid-Rich Microalgea in Wastewater for Biodiesel Production 1
Paul C. Kyriacopulos, Jason Ouellette, Ryan Leach, Ihab H. Farag

Seaing Green 9
Thaddaeus Andrew Webster, Benjamin Chiang, Ihab H. Farag

Simulation of Algal Growth In a Photobioreactor 18
Nadia J. Abunasser, Steven O. Salley, John Wilson, K. Y. Simon Ng

Algae-Based Hydrocarbon Production 19
Waqas Khatri, Steve Gabauer, Joeseph Chappell, Wayne Curtis

Conversion of Microalgae to Biodiesel at Supercritical Conditions 21
Prafulla Patil, Shuguang Deng

Oil Accumulation by Sewage-Grown Oleaginous Microorganisms 22
Andrew T. Ford, Dr. William T. French, Rafael Hernandez, William E. Holmes, Earl G. Alley

Experimental and Modelling Studies of the Bioconversion of Glycerol to Succinic Acid by Actinobacillus Succinogenes 23
Anestis Vlysidis, ChEnyu Du, Colin Webb, Constantinos Theodoropoulos

Impacts of New Technologies and Policies on Biofuels Production and Trade 31
Bhima Sastri, Audrey Lee, Thomas Alfstad, Richard L. Bain

Methane Gas Hydrate Equilibrium Inside Cylindrical Pores and Its Implications for Gas Production 58
Patricia Taboada-Serrano, Costas Tsouris

Novel Algal Photobioreactor Design and Operation for the Biofuels Industry 59
Nadia J. Abunasser, Mario Enrique Danton Garcia Perez, Haiying Tang, Steven O. Salley, John Wilson, K. Y. Simon Ng

Incorporation of Reaction Kinetics into a Multiphase, Hydrodynamic Model of a Fischer Tropsch Slurry Bubble Column Reactor 60
Donna Post Guillen, Anastasia M. Gribik, Steven P. Antal, Jonathan K. Shelley

Oleaginous Microorganisms Utilizing Glycerol as Sole Carbon Source for the Production of Lipids for Biofuels 75
J. Mathew Thomas, Rafael Hernandez, Todd French, William Holmes, Earl Alley, Jaclyn Hall

Global View of Technical Aspects of Fuel Ethanol Production 76
Charles D. Tereck

Accelerating Sample Preparation in the Biofuel R&D by Automated High Output Technologies 77
Josef G. Schroer

Windfuels – Competitive Transportation Fuels from Wind Energy and Waste CO2 78
David Doty, Siddarth Shevgoor

Supercritcal Phase Fischer Tropsch Synthesis from Coal Derived Syngas 86
Jeremiah Benoit, Kanchan Mondal, Tomasz Wiltowski

Catalytic Decarboxylation of Pyrolytic Oils Model Compound on Ceria-Zirconia Mixed Metal Oxide 87
Vissanu Meeyoo, Pramoch Rangsunvigit

Economic and Environmental Analysis of Hydrogen Production Strategies 88
Wei Yuan, Susilpa Bommareddy, Norman E. Sammons Jr., Mario R. Eden

Experimental Investigation of Methane Gas Production from Methane Hydrate 89
Yue Zhou, Marco J. Castaldi, Tuncel Yegulalp

Sds Adsorption on Cyclopentane Hydrates at the Hydrate/liquid Interfaace 90
Chi Lo

Simulation and Optimization of the Process Used in Colombia for the Production of Biodiesel from Palm Oil: a Kinetic Analysis and An Economical Approach 91
Federico Velosa Orduz, Jorge Mario Gomez

Process Development to Reduce Minor Component Content In Biodiesel 107
Haiying Tang, Rhet Joseph De Guzman, Steven O. Salley, K. Y. Simon Ng

Study of Antioxidants on the Oxidative Stability of Biodiesel 108
Haiying Tang, Rhet Joseph De Guzman, Steven O. Salley, K. Y. Simon Ng

Catalytic Decarboxylation of Oleic Acid Over Ceria-Zirconia Catalysts 109
Boonyarach Kitiyanan, Charothon Ung-jinda, Vissanu Meeyoo, Thirasak Rirksomboon, Pramoch Rangsunvigit

Product and Process Effects of Alternative Solid Fuels Usage at a Portland Cement Manufacturing Site 113
Steve R. Duke, Srikanth Akkapeddi, Dustin Swart, Anton Schindler, Don Stafford

Biodiesel Process and Control 114
Joseph Pearson, Simon Huang, Danah Hashem

The Future of Biodiesel as a Viable Alternative Fuel: Challenges and Opportunities 123
K. Y. Simon Ng, Steven O. Salley, John Wilson, Haiying Tang, Shuli Yan, Nadia J. Abunasser, Manhoe Kim

Electrical Characterization of Lead-Acid Battery at Load for Hev Applications 124
Wenhua H. Zhu, Robert U. Payne, Ying Zhu, Bruce J. Tatarchuk

Addressing Challenges In the Simulation of Lithium-Ion Battery Models 127
Vijayasekaran Boovaragavan, Venkat Subramanian

Robust Surrogate Modeling of Lithium Ion Batteries 129
Rajeswari Chandrasekaran, Jonathan Denis, Michael Skinner, Thomas F. Fuller

Impact of Ultra-High Carbon Surface Areas on Capacitory and Battery Performance 138
Galen J. Suppes, Bryan Sawyer

Silicon Nanotube. A Promising Candidate for Hydrogen Storage 139
Dapeng Cao, Jianhui Lan, Wenchuan Wang

Dynamics of Ti-Al-H Compounds In Ti-Doped NaAlH4: A DFT-MD Stability and Transport Study 140
Gopi Krishna Phani Dathara, Daniela S. Mainardi

New Insights into the Mechanism of Hydrolysis of Sodium Borohydride 141
Amy M. Beaird, Eyma Y. Marrero-Alfonso, Thomas A. Davis, Michael A. Matthews

Material Property Study of Thermal Battery Components 142
Babatunde Ekundayo, Irvin W. Osborne-Lee, Sean Walker

Density Functional Theory Studies on a Reversible Hydrogen Storage "Li-Mg-B-N-H" System 144
Nianthrini Balakrishnan, Pabitra Choudhury, Venkat Bhethanabotla, Babu Joseph

Investigation of Esterification of Biomass Pyrolysis Acids Over Sulfonic Acid-Functionalized Mesoporous Silicas 148
Shaojun Miao, Brent H. Shanks

Oligomerization of Propionaldehyde on Novel Basic Catalysts 149
Q. V. Amen, T. Sooknoi, L.L. Lobban, R.G. Mallinson, Daniel E. Resasco

Comparative Analysis of Biodiesel Production from Waste Cooking Oil Using Ferric Sulfate and Sulfuric Acid 156
Prafulla Patil, Shuguang Deng

Impact of Oxygen Incorporation In Molecular Structure on Pyrolytic Soot Formation and Oxidizability 157
Steven P. Crossley, Maria Pilar Ruiz, Liang Zhang, Daniel E. Resasco

Conversion of Acetone to Hydrocarbon on Titania Based Catalysts 158
Prashant Reuben Daggolu, Mark G. White

Deoxygenation of Model Biofuel Compounds Over Cs-NaX Zeolite Catalyst 159
Tanate Danuthai, Maria A. Peralta, Tawan Sooknoi, Somchai Osuwan, Lance L. Lobban, Richard G. Mallinson, Daniel E. Resasco

Catalytic Deoxygenation of Stearic and Oleic Acids Over Pd/C 167
Jeremy G. Immer, M. Jason Kelly, Benjamin D. Estabrook, H. Henry Lamb

Catalytic Hydrodeoxygenation of Liquid Biomass for Hydrocarbon Fuels 168
Kyung C. Kwon, Howard Mayfield, Theordore Marolla, Michael Mashburn

Gas Phase Processing of Sorbitol 179
Galen J. Suppes, Wei Yan

Conversion of Biomass-Derived Polyols Over Carbon Supported Pt-Re Catalysts 180
Edward L. Kunkes, Dante A. Simonetti, Ryan M. West, James A. Dumesic

Catalytic Deoxygenation of Benzaldehyde Over Supported Gallium Catalysts 181
Artit Ausavasukhi, Daniel E. Resasco, Tawan Sooknoi

Steam Reforming of Different Fractions of Biomass Pyrolysis Liquids 190
Pedro J. Ortiz-Toral, Justinus A. Satrio, Robert C. Brown, Brent H. Shanks

Millisecond Autothermal Catalytic Partial Oxidation of Glycerol to Syngas and ChEmicals 191
David Rennard, Lanny D. Schmidt

A Comprehensive Look at the Acidic Hydrolysis of Loblolly Pine Sawdust 192
Teresita Marzialetti, Mariefel Valenzuela-Olarte, Carsten Sievers, Travis Hoskins, Pradeep K. Agrawal, Christopher W. Jones

Biogas: Sludge to Biogas 193
Chikezie N. Nwaoha

Product Distribution from Fast Pyrolysis of Polysaccharides 194
Pushkaraj R. Patwardhan, Justinus A. Satrio, Brent H. Shanks, Robert C. Brown

Catalytic Gasification of Biomass in Supercritical Water 195
Fernando L. P. Resende, Phillip E. Savage

Investigation of High Temperature Gasification of Biomass In An Aerosol Flow Reactor 196
R. Bryan Woodruff, Christopher Perkins, Alan W. Weimer

Biomass Gasification: Catalytic Tar Removal Using Ni-Zeolite Catalysts 197
Prashanth R. Buchireddy, R. Mark Bricka

Low-Temperature Gasification of Cellulosic Biomass 198
Sameer Parvathikar, Johannes W. Schwank

Mechanism of Thermal Decomposition of Furan 199
AnGayle K. Vasiliou, Hans-Heinrich Carstensen, Mark R. Nimlos, John W. Daily, G. Barney Ellison

Author Index

Growing Lipid-Rich Microalgae in Wastewater for Biodiesel Production

Paul C Kyriacopulos, Chemical Engineering, University of New Hampshire (UNH) Durham, NH
pcw6@cisunix.unh.edu
Jason Ouellette, Biology, University of New Hampshire (UNH) Durham, NH
jason.ouellette@gmail.com
Ryan Leach, Chemical Engineering, University of New Hampshire (UNH) Durham, NH
Rax5@unh.edu
Ihab H. Farag, Chemical Engineering, University of New Hampshire (UNH) Durham, NH
03824-3591, 603-862-2313, ihab.farag@unh.edu

1- Introduction

1-1 Background. The public is becoming more aware of the need for alternatives to petroleum based fuels. The price of oil is increasing faster than new technologies, such as gas-electric hybrids, can compensate for. Biofuels such as biodiesel provide an environmentally safer alternative to petroleum fuels.

1-2 Biodiesel. Biodiesel is a renewable alternative to diesel fuel. It is made by the transesterification reaction of any lipid/oil with alcohol. In this process, three Fatty Acids which are bonded to a Glycerin group react with an alcohol to produce the biodiesel. The Glycerin, which is a byproduct of the reaction, could then be used in other products.

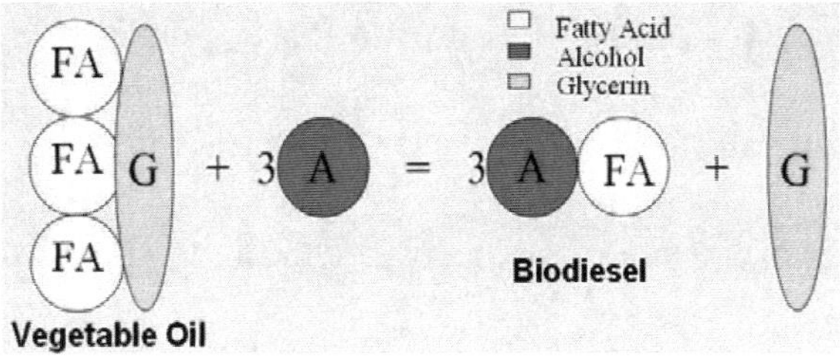

Biodiesel can be burned in existing diesel engines without any additional modifications. Also, biodiesel can be blended in any proportion with petroleum diesel. Potentially, biodiesel can be produced sustainably. It reduces the dependence on fossil fuels, produces far less carbon dioxide (CO_2) and other greenhouse gases (GHGs) than petroleum diesel, reduces other forms of air pollution such as carbon monoxide (CO) and sulfur dioxide (SO_2). Biodiesel has many advantages: It is a high-quality fuel; it can be immediately used within the existing infrastructure, and can support local agriculture and economic development.

1-3 Food Crop Challenges. Currently, biofuels such as biodiesel are produced from food plants such as soybeans and canola. Most plants are highly inefficient in utilizing the sunlight, using only 0.5% of the incident light. Plants are not very efficient in using the available land space. For example, soybean plants produce roughly 50 gallons of biodiesel per acre per year and Canola plants produce 90 gallons of biodiesel per acre per year. Huge amounts of land would be needed to produce enough biodiesel to meet the current US demands for diesel (roughly 60 billion gallons of diesel per year). It would take over one billion acres of land used for growing soybeans to produce enough biodiesel for the United States. This strategy causes deforestation, increases food prices as corn is being used as an "energy crop" to produce oil and ethanol instead of being used as food crop, and farmers are growing more corn to supply demand for biofuels.

1-4 Microalgae as a Feedstock. The term "Algae" is used to cover a wide range of organisms. Algae may be small, single-celled organisms to multi-cellular organisms which are usually found in water bodies. They are similar to plants in that they require the same three components to grow: sunlight, water, and carbon dioxide (CO_2). Just like plants, algae undergo the biochemical process of photosynthesis in which the algae use CO_2 and sunlight to store chemical energy and release oxygen. Swamps and salt lakes are natural sources of large amounts of algae. Compared with plants, microalgae tend to have better solar efficiency utilization. This is because they can be grown in continuous culture and hence provide higher yearly productivity. It is also believed that the addition of CO_2 to microalgae is simpler than the case of plant crops. From the point of view of producing biodiesel, the most important constituents of microalgae are lipids and fatty acids. These provide the source of energy for the microalgae. Earlier research indicated that the lipids content of algae ranges from 2% to 40% based on the dry mass of algae.

Microalgae have the potential to solve many of the current problems with biofuels. They grow faster than any food crop, and can produce between 5,000 and 15,000 gallons per acre. That is about 100 times the biodiesel yield of soybeans. The left-over algae biomass (after oil extraction) can be fermented to alcohol, or can be used as a source of protein in the cattle feed. Based on an average of 10,000 gallons of biodiesel per acre, we would need 6 million acres of land containing algae to produce enough biodiesel for the United States, which would be close to the size of New Hampshire (assuming all of New Hampshire to be agricultural land). Algae also have the ability to grow in closed bioreactors, which can be situated in areas that are unfit for farming. This is attractive both on a financial standpoint and an ethical standpoint.

1-5 Photobioreactors. These are closed vessels where the microalgae can grow under optimum conditions without the concern about contamination which could consume or out-compete the selected strain of microalgae. A photobioreactor can be as simple as a lab scale conical transparent glass beaker exposed to light, to a more sophisticated industrial scale design. The challenge in photobioreactors is to be able to achieve high growth rate of the microalgae in a cost-effective manner.

1-6 Nutrients. Algae use light energy (e.g., sunlight) to obtain their inorganic compound's nutritional needs. The three key nutrients that influence the growth are nitrogen, phosphorous and carbon. The metabolism of algae is strongly affected by limitation in nutrients. It is well known that most algae stressed by nitrogen limitation would increase their lipid content. Phosphorous is an important nutrient. Phosphorous limitation would slow down the growth rate. Carbon is the only source of photosynthesis. When running algae growth experiments, it is necessary to use soluble nutrients for the algae to grow. The most important nutrients are nitrates and phosphates followed by sodium chloride and silicate. A typical nutrient medium solution includes macro nutrients (sodium, phosphates and nitrates), micronutrients (iron, sodium, and Zn), and vitamins.

1-7 Wastewater Use Motivation. Municipal wastewater usually contains nitrogen, phosphorous and other nutrients. This may be a desirable environment for growing algae. It was hypothesized that the substances dissolved in the wastewater and effluent would serve as nutrients and support microalgae growth and lipid production. The UNH Biodiesel Group is investigating the use of municipal wastewater to grow lipid-producing algae. These algae would then be used as a feedstock for biodiesel production.

Wastewater is an attractive medium because it is inexpensive compared to chemically defined media, and is readily available wherever there is a municipal wastewater treatment facility. If wastewater proves to be an effective medium, it may be possible to build algae biodiesel production facilities that would work directly off the effluent from wastewater treatment facilities. This wastewater algae biodiesel technology will not place additional demand on freshwater supplies needed for domestic, industrial, and agricultural use.

2- Goal

The economics of producing biodiesel from algae requires the most effective utilization of the algae. The goal of this project is to improve the biodiesel process economics by using microalgae-produced lipid/oil. This is attempted by using wastewater to grow the microalgae.

3-Experimental Methods

As discussed, algae require sunlight, water, and carbon dioxide (CO_2) to grow. The light was supplied by an array of fluorescent lights. CO_2 was supplied by bubbling air into the algae solution. The water used was obtained from the Durham Wastewater Treatment Facility in Durham, New Hampshire. Water samples used in our experiments included homogeneous untreated waste water, dechlorinated effluent (simply termed effluent) , which was a treated wastewater suitable to be released into the environment, and finally a mixture of 50% wastewater and 50% dechlorinated effluent. Our photobioreactor was a two-liter clear glass conical flask in which the algae and water solution were placed. Air was bubbled into the solution to provide a source of CO_2 and to stir the algae solution. The clear flask was placed in front of fluorescent light. All experiments were done indoors. The work proceeded in two steps.

3-1 Step 1. Wastewater Contamination. Initially, the untreated wastewater was observed for unwanted growth in the system. The solutions tested were wastewater alone (as a blank), wastewater plus algae, and finally wastewater plus algae and 0.1 Molar NaCl. It could be seen that there was contamination in the wastewater and this contamination would most likely lead to negative competition between the lipid-producing microalgae (which is highly desirable) and a contaminating algae or bacteria. Thus, the untreated wastewater would not be a desirable medium for algae growth unless the contaminating organisms could be controlled.

3-2 Step 2. Effect of Solution Media. The microalgae were grown in three solutions: a nutrient solution containing 0.1 Molar NaCl (termed Nutrient Solution in the Figures), a dechlorinated effluent solution containing 0.1 Molar NaCl (termed Effluent in the Figures), and finally the mixture of 50% effluent and 50% wastewater plus 0.1 Molar NaCl (termed Effluent/ Wastewater Blend in the Figures). The outcomes are that the nutrient solution produced the algae with the highest oil content, while the dechlorinated effluent showed a good promise for algae growth.

3-3 Monitoring Algae Growth Rate. In order to reach the above outcomes, the microalgae strain was tested at room temperature for two parameters over a time period of about 7-14 days per trial. The first parameter was growth rate, which was determined by measuring the solution absorbance using a Bausch & Lomb Spectronic 20 spectrophotometer. Absorbance is used to measure turbidity, and higher absorbance reading indicated higher algae concentration.

3-4 Monitoring Algae Lipid Concentration. The second parameter is the lipid/oil concentration in the microalgae. This was found using Cooksey's method. All solutions were first normalized to an equal turbidity then Nile Red dye was added to stain the lipids in the algae. The solution was then placed in aVarian SF-330 spectrofluorometer to measure the fluorescence of the dyed lipids. When recording the data from the spectrofluorometer, it was observed that the fluorescence (of the normalized sample) gradually increased to a maximum value then decreased. The data was recorded in 5 minute increments. When the data was analyzed, the highest and lowest fluorescence values were subtracted from each other to find the peak normalized fluorescence of the sample. It is assumed that the peak normalized fluorescence is proportional to the algae lipid concentration.

4- Results

Figure 4.1 displays the algae cell concentration (as measured by the absorbance/turbidity of the solution) vs time for the three media. Clearly, the best algae growth takes place in the nutrient solution. Some growth took place in the effluent/ wastewater blend. The growth of algae in the effluent is encouraging.

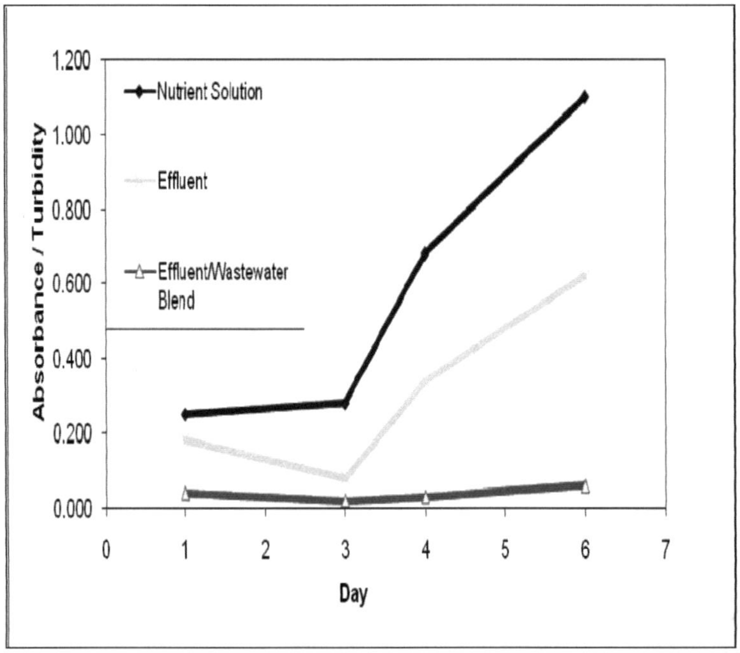

Figure 4.1 Effect of solution media on algae concentration.

Figure 4.2 displays the algae lipid concentration (as measured by the peak normalized fluorescence) vs time. The algae oil concentration in the effluent media and the effluent/ wastewater blend had about 20% of the nutrient solution.

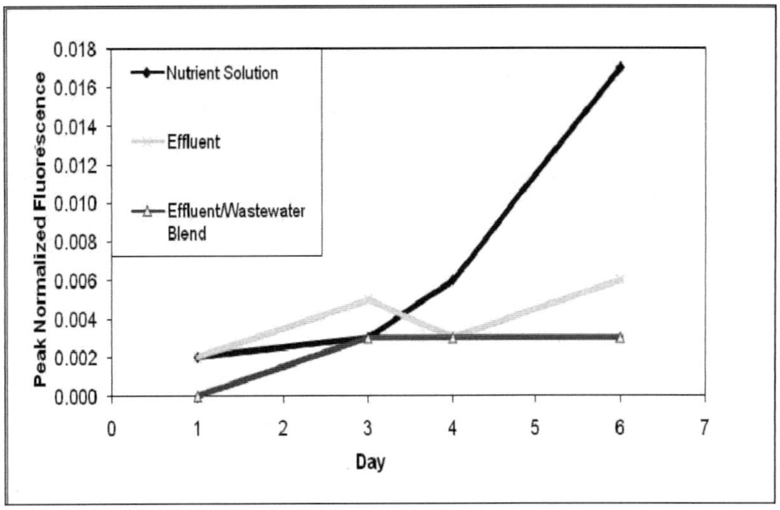

Figure 4.2 Effect of solution media on oil concentration in algae.

The oil content of the algae solution was obtained by multiplying the absorbance/turbidity (algae concentration in the solution) times the peak normalized fluorescence (lipid concentration in algae). The oil content is plotted vs time in Figure 4.3. The results show that the mixture of 50% effluent and 50% wastewater had an oil content of about 5% that of the nutrient solution, while the dechlorinated effluent was about 19.9% that of the nutrient solution.

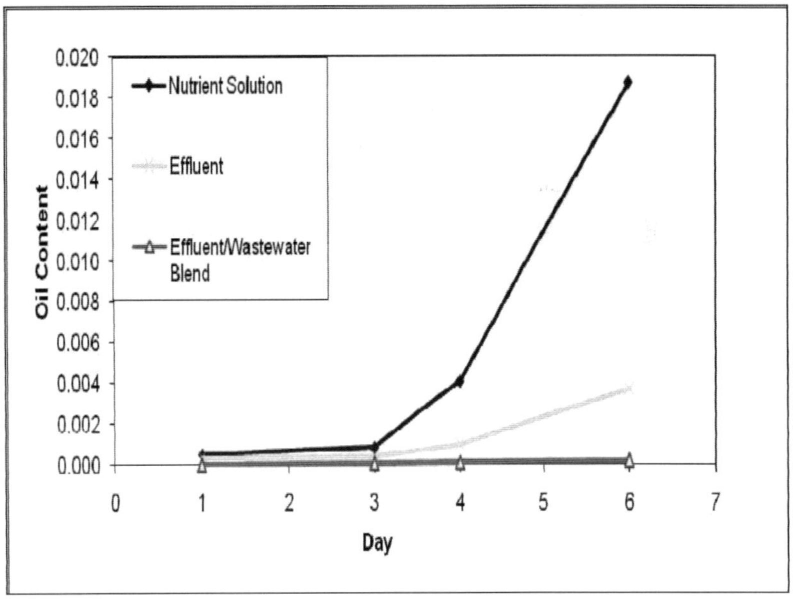

Figure 4.3 Effect of solution media on oil content of algae solution.

After testing the algae for oil content, the operation of the photobioreactor had to be adjusted. After 3-4 days of algae growth, approximately half of the solution had evaporated. To help in decreasing this evaporation, which was exacerbated by the air being bubbled in to mix the algae, the air flow rate was lowered. Also, various reactor designs were produced by other researchers in the UNH Biodiesel Group. Although these were successful in producing algae, contamination was still a major problem. New photobioreactors are still being developed.

5- Conclusions

Microalgae grown in nutrient and salt solution produced the greatest oil yield while the microalgae grown in effluent showed good promise. More tests will be taken with the various wastewaters to determine the best method of growing algae for the conversion to biodiesel. These results will be used in future planning for microalgal biodiesel research.

6- Future Plans

New photobioreactor designs will be tested in order to find a better system. During the experiment, it was noted that the wastewater had contaminants which may affect algae growth. To prevent biological contamination, ultraviolet irradiation of the wastewater will be studied..

Acknowledgment:

The authors would like to acknowledge the support of the UNH Hamel Center for Undergraduate Research and the UNH Chemical Engineering Department.

References:

Cooksey, K.E., Guckert, J.B., Williams, S.A., and Callis, P.R. (1987) Fluorometric determination of the neutral lipid content of microalgal cells using Nile Red. *Journal of Microbiological Methods, 6*, 333-345.

Farag, I.H. (2006) "Biodiesel, The Better Air Quality Alternative to Diesel," EST BAQ (Better Air Quality) Conference Dec 13-15, 2006, Yogjakarta, Indonesia.

Farag, I.H. (2007) "Biodiesel, The Renewable Alternative to Diesel," 7th Int. Conference & Exhibition on Chemistry in Industry CHEMINDIX 2007, March 23-25, 2007, Manama, Bahrain

Farag, I.H. (2008), "Biodiesel: Challenges and new Fronts", Keynote Paper, International Biodiesel Workshop, National Research Center, March 23-25, 2008, Cairo, Egypt

SEAING GREEN

Benjamin Chiang, Thaddaeus Webster, Ihab H. Farag,
Chemical Engineering, University of New Hampshire (UNH) Durham, W315 Kingsbury Hall,
33 College Rd, NH 03824-3591, 603-862-2313, ihab.farag@unh.edu

1- Background

1.1 Oil Consumption in the U.S. The U.S. is one of the world's largest importers of oil. Currently the U.S. has an annual diesel demand of 65 billion gallons. The increase in oil prices provided a strong incentive to research renewable fuel sources. One research area is the area of biofuels, expected to create local jobs and offer alternatives to the U.S. reliance on petroleum based fuels.

1.2 Biofuels Challenges. Currently biofuels are being produced from corn, soy bean, canola, and sugar cane. While these fuel sources are renewable and are more environmentally friendly than petroleum fuel sources, they have their draw backs. First and foremost these crops are already being used as "food crops" for the U.S. and the rest of the world. The use of these as "energy crops" has led increase in the price of food. In addition, Plant energy crops are a dispersed source of energy requiring large land acreage to produce the required oil feedstock. For example an acre of soy beans only produces about 90 gallons of useable oil per year.

1.3 Biodiesel from Algae. One potential source of biodiesel production is to grow single celled high lipid algae. Depending on the microalgae used, the lipid content may be 20-40% (dry mass basis). The high lipid content makes these algae suitable for biodiesel production. It has been hypothesized that an acre of algae could produce anywhere from 5000-15000 gallons every year, which is much higher than traditional crops. Algae utilize photosynthesis just like plants do to store carbon and make lipids, and grow very rapidly. Algae can be grown in almost any conditions as long as they are exposed to sunlight, a nutrient solution, and air (more importantly carbon dioxide or CO2). In addition algae do not require soil for growth and can be located on marginal land, thus avoiding the competition of land use for food crops.

1.4 Photo-bioreactors. A photo-bioreactor is a reactor in which algae is grown. It provides a light source (sunlight or man-made), nutrients, and air to help grow the organisms. When algae were first studied for use as a feed stock for biofuels, they were grown in open system (ponds) photo-bioreactors in which the algae were exposed to the environment. These open system photo-bioreactors suffered contamination. In addition, it was very hard to control many other important variables such as temperature and pH.

In recent years many researchers have been looking at enclosed photo-bioreactors in which variables like temperature, contamination, and pH can more easily be controlled, thus making it much easier to grow a strain of algae than in the open system photo-bioreactors.

1.5 Nutrients. Soluble nutrients are needed for the algae to grow. The most important nutrients are nitrates and phosphates followed by sodium chloride and silicate. A typical nutrient medium solution includes macro nutrients (sodium, phosphates, nitrates, and silicate), micronutrients (iron, sodium EDTA, Co, Zn, copper sulfate, and Mn), and vitamins (B7 and B1).

2. Goal/Objectives

The overall goal of this research project is to find a more economical technique of growing algae that not only grows the algae quickly, but also allows the algae to have high oil yields. The objective of this research project was to study the effect of salinity on the growth and oil production of two algae species X1 and X2.

3. Salinity Experiments

Saltwater was used a nutrient solution to grow algae for two reasons. First it is a very abundant resource. Growing algae in saltwater would save potable water used to make nutrient solutions.

Also the increased salt concentration would probably kill most of the contaminants, which probably are not adapted to the salinity. The testing was done in three phases.

3.1 Phase One. The two algae strains X1 and X2 were placed in a regular nutrient solution (the control), and a 0.1 M sodium chloride and nutrient solution. This was done to determine whether an increase in the amount of salt would have positive effects on the algae growth and oil content. During a two weeks growth period, the algae and lipid concentration were monitored.

3.2 Phase Two. The two algae strains were placed in pure ocean water. This was done for two reasons. First, the results from phase one indicated that the algae had grown better and produced more oil when the concentration of sodium chloride increased. Second, if the algae could in fact grow better in pure salt water than in nutrient solutions, then the process of growing the algae would be much more economical. At the end of phase two it was determined that the algae could not grow in pure salt water, because it lacked the nutrients that the algae needed to survive.

3.3 Phase Three. The algae were grown in mixture of 50% (by volume) ocean water and 50% 0.1 M sodium chloride and nutrient solution, vs. the 0.1 M sodium chloride and nutrient solution. The objective was to determine the effects a mixture of nutrient solution and pure ocean water would have on the oil content and growth of the two algae strains being investigated, X1 and X2.

4. Measurement Methods

4.1 Measuring Turbidity/Absorbance. Two characteristics are employed when screening algae species as a viable source for biodiesel. These are the algae growth rate, and the amount of lipid produced. The algae growth rate is determined by measuring the algae concentration over time. The turbidity/absorbance of the algae solution is directly proportional to the algae concentration. By measuring the turbidity over a period of time (7 – 14 days in most experiments) it is possible to determine the relative growth rate of different algae species. The algae solutions turbidity of was measured using a Bausch & Lomb Spectronic 20 spectrophotometer at 682 nanometers.

4.2 Algae lipid content. While it is important to select fast growing algae for the production of biodiesel, it is just as important to find high lipid algae. When extracted, these lipids can then be turned into biodiesel. To measure the change in the lipid content of a given algae species over a period of time, the method of Cooksey et.al. was used in which the oil is stained with Nile Red due and the fluorescence is measured. A Varian SF-330 spectroflourometer with a Xenon Lamp Power Supply was used to measure the algae fluorescence. Algae cultures were normalized (diluted) to an equal turbidity. By measuring the peak fluorescence over a period of time it was possible to determine how the lipid content of the algae changed.

5-Results

5.1 Phase One: Fresh water vs. Salt water results. These experiments involved comparing algae grown in a regular nutrient solution (denoted by X2 nutrient in the figures) and a 0.1 M sodium chloride and nutrient solution. (denoted by X1 Sea and X2 Sea in the figures). Figure 1 shows the turbidity (which measures algae cell concentration).

Figure 2 shows the peak normalized fluorescence; a measure of the lipid concentration in the algae cells.

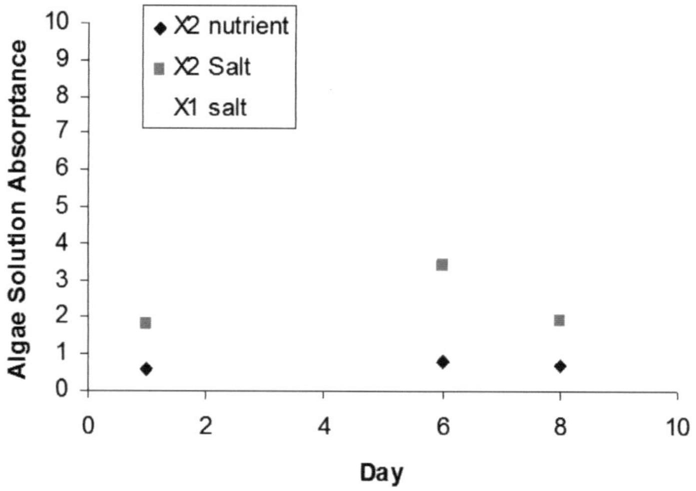

Figure 1: Effect of salt on algae growth

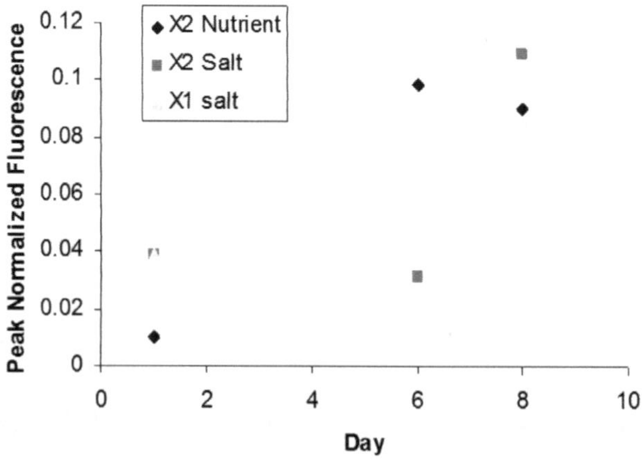

Figure 2: Effect of salt on algae lipid concentration

Figure 3 is the product of the turbidity (Figure 1) and the peak normalized fluorescence (Figure 2). It is a measure of the lipid content of the algae solution. The graphs show that algae species X1 grew faster than algae species X2 (Fig. 1), and that X1 had a higher fluorescence (Fig 2) and a higher lipid content (Fig. 3).

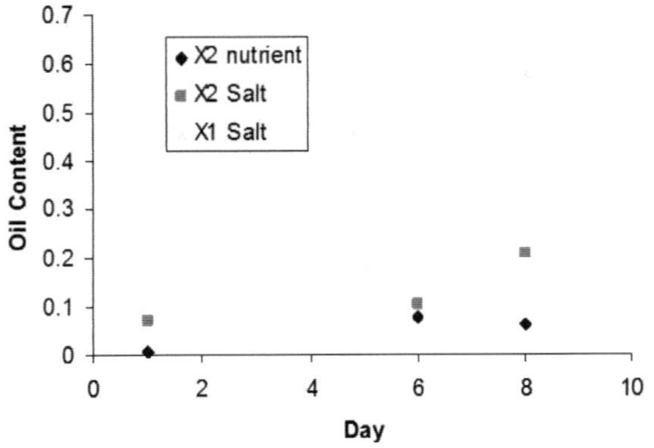

Figure 3: Effect of salt on algae lipid content

5.2 Phase Two: growing the algae in pure ocean water The second set of experiments involved growing the algae in pure ocean water. Algae species X2 was grown in a normal nutrient solution, a 50% sea water and 50% normal nutrient solution, and a pure ocean water solution. Figure 1 shows the turbidity (which measures algae

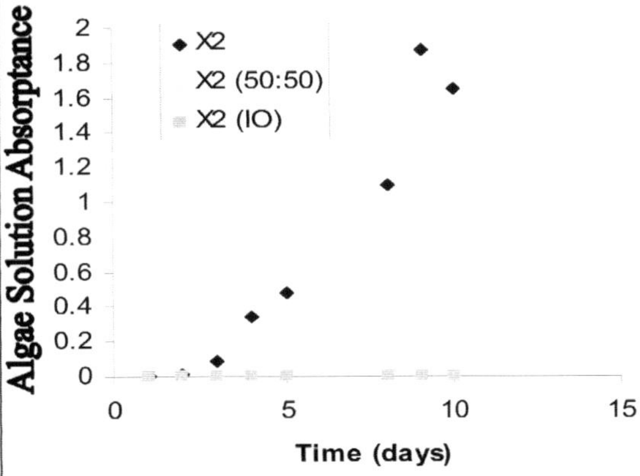

Figure 4: Effect of pure ocean water on algae growth

<table>
<tr>
<td>cell concentration).

Figure 5 shows the peak normalized fluorescence; a measure of the lipid concentration in the algae cells. Figure 6 is the product of the turbidity (Figure 1) and the peak normalized fluorescence (Figure 2). It is a measure of the lipid content of the algae solution. The graphs show that algae species. These results show that the algae do not grow well in the pure ocean water (possibly due to lack of nutrients) but they do grow well in the 50% sea water and 50% normal nutrient solution

.</td>
<td>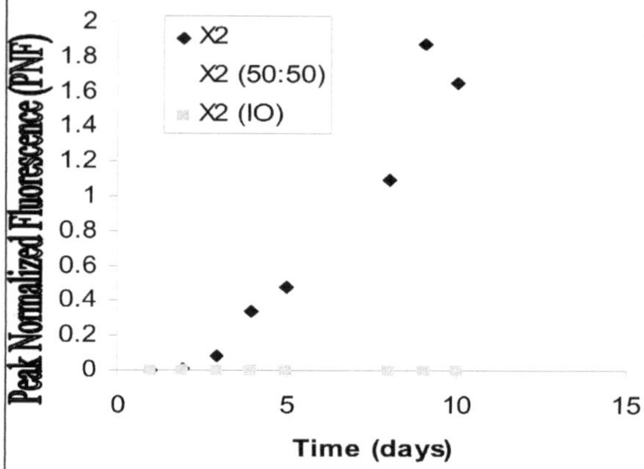

Figure 5: Effect of pure ocean water on algae lipid concentration
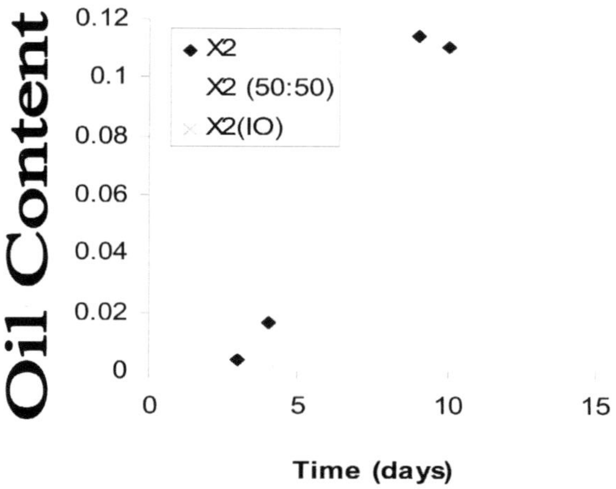

Figure 6: Effect of pure ocean water on algae lipid content</td>
</tr>
</table>

5.3 50% sea water and 50% normal nutrient solution vs. normal nutrient solution

The third set of experiments involved comparing algae grown in a normal (0.1 M NaCl) nutrient solution (denoted by X1 and X2 in the figures) versus the growing the same X1 and X2 algae in a mixture of 50% sea water and 50% normal nutrient solution. This is denoted X1 Sea and X2 Sea in the figures. Figure 7 shows the turbidity (which measures algae cell concentration). Figure 8 shows the peak normalized fluorescence; a measure of the lipid concentration in the algae cells. Figure 9 is the product of the turbidity and the peak normalized fluorescence. It is a measure of the oil content of the algae solution. From the above graphs it was again observed that algae species X2 grew faster than algae species X1 (figure four), but from figures five and six it was clear that algae species X1 had a higher fluorescence and a higher oil content in both the 50:50 mixture of ocean water and 0.1 M sodium chloride nutrient solution and the 0.1 M sodium chloride nutrient solution than algae

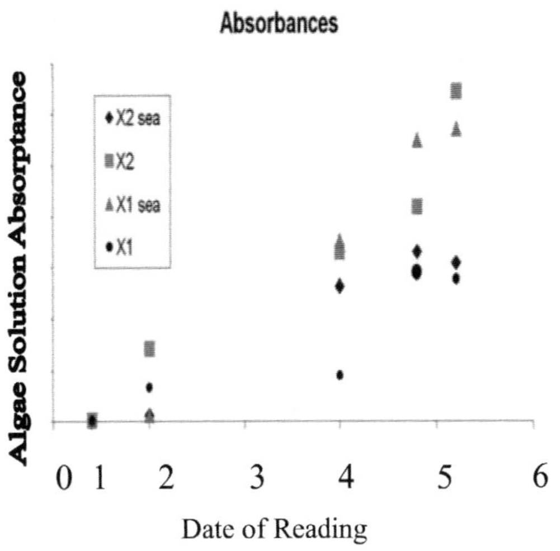

Figure 7: Effect of 50% sea water and 50% normal nutrient solution on algae growth

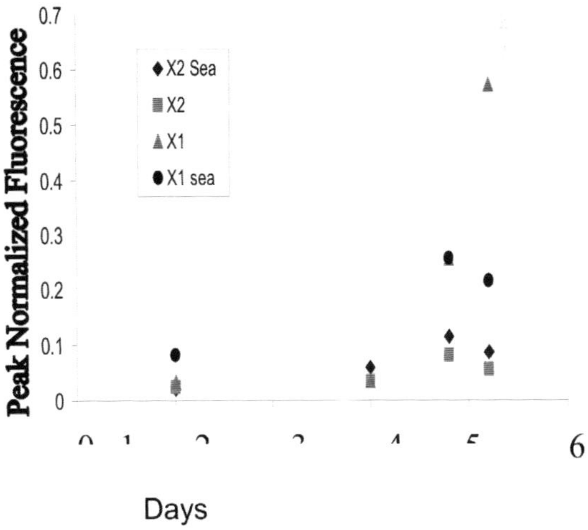

Figure 8: Effect of 50% sea water and 50% normal nutrient solution on algae lipid concentration

species X2	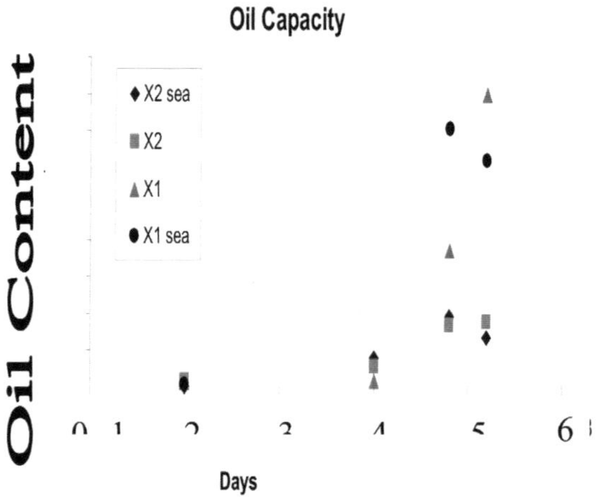 Figure 9: Effect of 50% sea water and 50% normal nutrient solution on algae lipid content

5-Conclusions:

The purpose of this research project was to determine the viability of using ocean water as a possible nutrient source to grow the two freshwater algae species that were studied. While the results of the screening process for experiment show that algae species X1 performed better than algae species X2, these results are qualitative, and more experiments must be done to determine whether the trends that were observed in these experiments can be replicated. The results of experiment number two helped us to realize that pure ocean water can not be used as a sole nutrient source, because it lacks a lot of crucial nutrients that algae, both freshwater and saltwater, need in order to grow and flourish. Even more promising were the results seen in experiment three in which algae species X1 was able to grow rather well in the 50% sea water and 50% normal nutrient solution thus making the use of ocean water as a nutrient source a possibility.

Also based on the above results the UNH Biodiesel program plans to begin performing research with salt water algae in an attempt to find a species that can either grow well in pure ocean water, or in a nutrient solution in which ocean water is the stock source, instead of using potable water. We are very excited to continue researching growing algae in the hopes that it can be used as an alternate feedstock to produce biodiesel.

Acknowledgment:

The authors would like to acknowledge the support of the UNH Hamel Center for Undergraduate Research and the UNH Chemical Engineering Department.

202c Simulation of Algal Growth In a Photobioreactor

Nadia J. Abunasser[1], Steven O. Salley[2], John Wilson[3], and K. Y. Simon Ng[2]. (1) Chemical Engineering and Materials Science, Wayne State University, 461 Burroughs, Detroit, MI 48202, (2) Department of Chemical Engineering and Materials Science, Wayne State University, 5050 Anthony Wayne Dr., Detroit, MI 48202, (3) National Biofuel Energy Laboratory, NextEnergy, 461 Burroughs ST, Detroit, MI 48202

If algae are to be considered as an economical alternative feedstock for biodiesel and other renewable fuels, an efficient cultivation method needs to be designed. In theory, all algae needs to grow are sunlight, carbon dioxide and some trace elements. One method of cultivation currently being considered is the photobioreactor. A complete mathematical model of the system would be useful to analyze and improve the productivity, efficiency and economics of the reactor. To date there is no such model of the growth kinetics as a function of light intensity which varies in the reactor both as a function of distance from the source and cell concentration. The objective of the current work is to model the algal growth in a novel external loop airlift reactor as a function of light intensity and shear stress to aide in the improvement of the design of the photobioreactor. The photosynthetic factory model proposed by Eilers and Peeters (1988) is used to determine which cells are photosynthetically active. The variation in the light intensity is modeled initially using the Beer-Lambert equation, and then modified using experimental data. The geometry of the system is also considered in the modeling of the growth kinetics. All the equations were solved using Matlab.

Algae-Based Hydrocarbon Production

Waqas Khatri[1], Steve Gabauer[2], Joeseph Chappell[3] and Wayne Curtis[2], (1)Industrial Engineering, The Pennsylvania State University, 232 Fenske Laboratory, University Park, PA 16802, (2)Chemical Engineering, The Pennsylvania State University, 232 Fenske Laboratory, University Park, PA 16802, (3)Department of Plant and Soil Sciences, University of Kentucky, 301B Plant Sciences Building, 1405 Veterans Drive, Lexington, KY 40546-0312

Abstract:

This presentation focuses on characterization of growth and hydrocarbon production from *Botryococcus braunii* as part of an overall project to develop the tools to migrate the unique metabolic capacity of this organism for the production of C30+ isoprene hydrocarbons to other algae and higher plants. This work compliments our ongoing photobioreactor design work to achieve maximum photon use efficiency at minimum power input for algae growth systems. Utilizing stoichiometrically designed growth medium and semi-continuous growth conditions, we demonstrate growth to cell concentrations greater than 20 grams dry weight per liter and producing more than 1.5 grams dry weight per liter per day. Cell density drops off rapidly at higher dilution rates, but biomass produced stays relatively constant. Under the high light conditions used in this work, there is a very significant increase in aggregate size as the periodic steady-state density drops, and this is reflected in an order-of-magnitude variation in optical density (OD_{550}) to dry weight ratio which is problematic for bioreactor monitoring. Strategies and progress on genetically engineering hydrocarbon synthesis pathways into *Chlamydomonas reinhardtii* will be noted but details will be presented in a poster session. Effects of nitrogen availability on productivity have also been examined. Oil content and composition data will be available shortly.

Background:

Algae are getting considerable attention for their ability to capture CO_2 emissions and turn them into biofuels. Instead of the typical approach to producing fatty acids for biodiesel production, *Botryococcus braunii* (**Figure 1**) grows in colonies that contain large quantities of isoprene hydrocarbons in their matrix. This study was undertaken to understand the control of this metabolic pathway under high-intensity growth conditions. Since the culture did not tolerate either high nutrients or high light very well, the cultures had to be 'adapted' using fed-batch techniques leading to subsequent semi-continuous production.

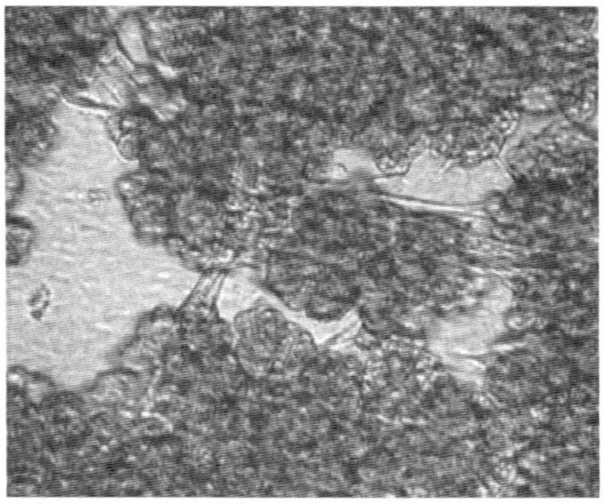

Figure 1: Photomicrograph of colony growth habit of *Botryococcus braunii* and surrounding isoprene matrix.

Results

Figure 2 shows a semi-continuous growth evaluation of *Botryococcus braunii* at various effective dilution rates where a proportion of the culture was drawn off and replaced every other day. Media was designed based on historic defined media for algae and plant cell culture to avoid accumulation of unused nutrients.

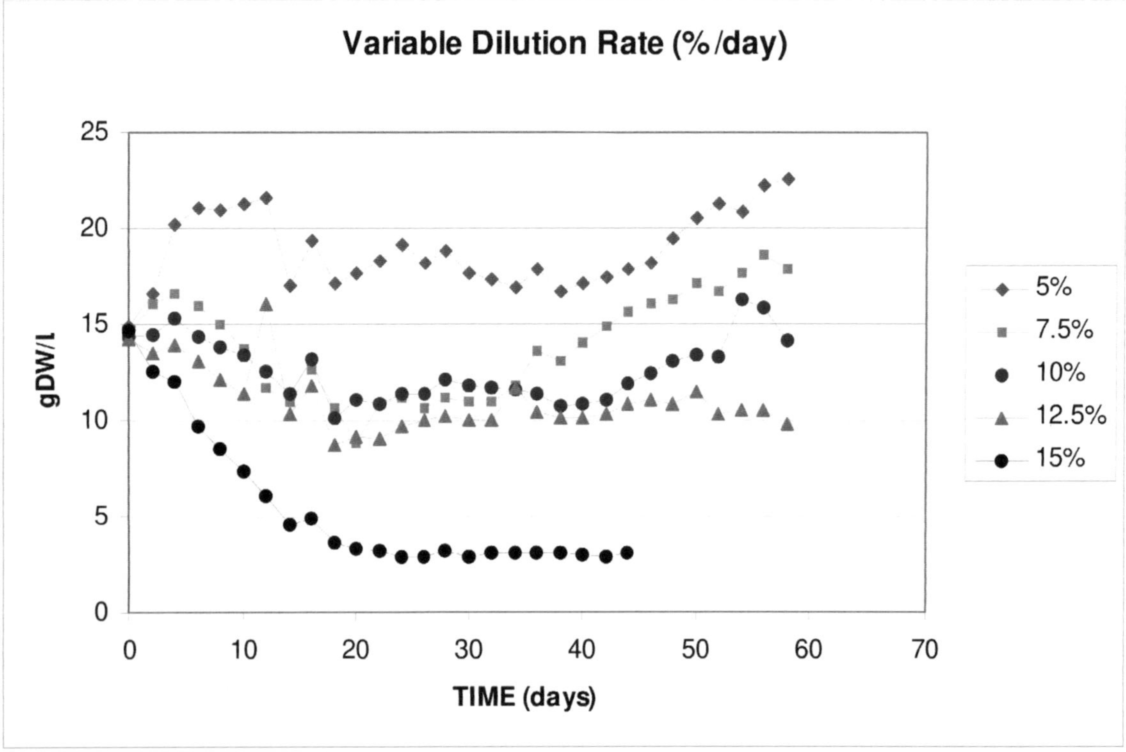

Figure 2: Growth of *B. braunii* in semi-continuous culture. Numerous modifications to media and nutrient balancing were implemented while gaining experience with long-term high-density culture.

Initial approximate methods of balancing nitrogen input based on measurements of dry weight and optical density were refined to reflect the large differences in optical density to dry weight ratio. At high dilution rates (near washout) the algae formed very large aggregates approach a millimeter in size. The large variation in culture dependent aggregate size results in a 3-fold variation in OD/DW ratio which is not observed in typical micro-algal culture.

Despite the decline in periodic steady-state cell density at different effective dilution rates, the volumetric productivity of the cultures was surprisingly constant reflecting a likely limitation by photon flux. Samples are currently being subjected to biochemical analysis including oil content and parameters to reflect biologically active biomass since a significant portion of the dry weight can be hydrocarbon & isoprene matrix.

202e Conversion of Microalgae to Biodiesel at Supercritical Conditions

Prafulla Patil, New Mexico State University, P.O. Box 30001, MSC 3805, Las Cruces, NM 88003-8001 and Shuguang Deng, Chemical Engineering Department, New Mexico State University, New Mexico State University, P.O. Box 30001, MSC 3805, Las Cruces, NM 88003.

Biodiesel derived from oil crops is a potential renewable and carbon neutral alternative to petroleum fuels. Unfortunately, biodiesel from oil crops, waste cooking oil and animal fat cannot realistically satisfy even a small fraction of the existing demand for transport fuels. As demonstrated in this work, microalgae appear to be a viable source of renewable biodiesel that is capable of meeting the global demand for transport fuels. Biodiesel from microalgae using supercritical methanol method and conventional alkali transesterification were studied in detail. A non-catalytic biodiesel production route with supercritical methanol has been developed; it allows a simple process and high yield because of simultaneous transesterification of triglycerides and methyl esterification of fatty acids. The most important variables affecting the methyl ester yield during the transesterification reaction are the molar ratio of alcohol to vegetable oil and the reaction temperature. The fuel properties of biodiesel produced were compared with ASTM standards for biodiesel and regular petroleum diesel.

202g Oil Accumulation by Sewage-Grown Oleaginous Microorganisms

Andrew T. Ford[1], Dr. William T. French[2], Rafael Hernandez[3], William E. Holmes[4], and Earl G. Alley[4]. (1) School of Chemical Engineering, Mississippi State University, 408 11th St. S, Mississippi State, MS 39759, (2) Chemical Engineering, Mississippi State University, Starkville, MS 39759, (3) Dave C. Swalm School of Chemical Engineering, Mississippi State University, 330 Swalm President's Circle Dr., Mississippi State, MS 39762, (4) Mississippi State Chemical Laboratory, Mississippi State University, Box 9595, Mississippi State University, MS 39762

The looming energy crisis and recent research in renewable fuels has led to an understanding that any solution will most likely involve many different strategies. The ability of certain microorganisms to accumulate oils utilizing sugars as substrates has long been known. In this paper we examine the potential for using a consortium culture of oleaginous microorganisms along with current infrastructure of wastewater treatment facilities to provide an inexpensive source for growth media. Previous work has demonstrated sewage wastewater to be an excellent source for inorganic nutrients. Using sewage supplemented with additional carbon sources, greater cell yields than those obtained using commercially available sources were acquired. Next this consortium of oleaginous microorganisms was offered sugars from acid hydrolyzates to begin triglyceride accumulation. Preliminary results show that at least 20% conversion rates of oil from sugar are possible using individual members of the consortium. Thorough examination of optimum sugar concentration and growth conditions is expected to increase these yields. The combination of low cost substrates using the infrastructure already in place (i.e. wastewater treatment facilities) and lignocellulosic sugars strengthens the potential to produce billions of gallons of renewable feedstock for the production of biodiesel or renewable diesel.

Experimental and Modelling Studies of the Bioconversion of Glycerol to Succinic Acid by Actinobacillus Succinogenes

*Anestis Vlysidis, Chenyu Du, Colin Webb and Constantinos Theodoropoulos**
School of Chemical Engineering and Analytical Science,
University of Manchester, Manchester, M60 1QD United Kingdom

Abstract

An unstructured kinetic model for the growth of *Actinobacillus succinogenes* on glycerol as the only carbon source is proposed in this study. The model describes cell growth considering both substrate and product inhibition. and its parameters are estimated by fitting the model predictions to experimental data. Substrate consumption and product formation rates are described by the Luedeking-Piret model. The main product of the process is succinic acid while by-products like acetate, formate and ethanol have very low concentrations. The main environmental factors that affect the bioprocess were examined and optimum conditions in terms of yield, final succinic acid concentration and productivity were evaluated by a factorial experimental procedure. The examined parameters were the supply of CO_2, the pH level and the redox balance. Experiments using different initial glycerol concentrations at the "optimum" environmental conditions were carried out and simulation studies were performed using the proposed model.

1) Introduction

Biodiesel is an alternative transport fuel which is made from renewable sources such as vegetable oils and animal fats. Recently, biodiesel production has received increasing attention due to national and international legislations, increasing petroleum prices, depletion of conventional fuels, concerns about food-to-fuels production and potential environmental benefits of biofuels. As the demand and the production of biodiesel grow fast, the development of methods to increase the sustainability of the biodiesel industry becomes an urgent topic.

Glycerol which is the main by-product of the biodiesel industry represents 10% (w/w) of the biodiesel produced and constitutes the basic bottleneck because of its currently limited applications due to the impurities that it contains [1]. In the last two decades, 1.3-propanediol has received increasing interest as a potential bulk chemical produced from crude glycerol since this bioconversion can achieve high final product concentrations and high productivities [2], [3]. The large potential that bio-processing exhibits, has instigated research efforts into further prospects of crude glycerol. Recently, fermentation studies of glycerol have examined the production of other products such as ethanol [4] and succinic acid.

Succinic acid is a C4 linear saturated dicarboxylic acid and is currently produced by petrochemical precursors through reduction of maleic anhydride and by microbial fermentation from glucose and/or other sugars [5]. Succinic acid has numerous applications and it can be used as surfactant, detergent, ion chelator and also in the food and pharmaceutical market. An extra advantage of the succinate bioprocess is that the microbial formation of succinic acid requires the fixation of a greenhouse gas: carbon dioxide [5].

Fermentation processes are affected by many environmental factors and usually produce a number of by-products. Research for the optimal environmental conditions in bio-

* Corresponding author: k.theodoropoulos@manchester.ac.uk

processes not only targets the maximisation of the desired product but also the minimisation of side products. From the environmental and physiological conditions that seem to affect the succinic acid production the most important are the amount of dissolved CO_2 in the fermented bed, the availability of electron donors, the pH values during the fermentation and the initial substrate concentration [6].

In particular, M. J. Van der Wert et al [6] have examined the effect of carbon dioxide on the succinic acid production on *Actinobacilus succinogenes* 130Z strain by supplying initially amounts of magnesium carbonate in the bioreactor. Results showed that there is a direct relationship between CO_2 and succinic acid production and simultaneously an inverse behaviour on the side-products formation (acetic acid, formic acid and ethanol). Furthermore, J B McKinlay et al [7] suggested the provision of hydrogen during the fermentation as electron donor to achieve a certain redox balance. According to the overall succinate reaction, four electrons $[H^+]$ are required to form succinic acid from glucose, thus electron supply in the form of hydrogen as gas during the fermentation resulted in a significantly higher succinic acid production.

Samuelov et al [8] studied the influence of pH and CO_2-HCO_3 levels on succinic acid production by using the bacterium *Anaerobiospirillum succiniciproducens*. They discovered that from neutral towards acidic pH (7.2-6.2) and from low to high amounts of CO_2 the succinic acid production was increased in favour of lactic acid and ethanol. The pH, during the succinate bioprocess, decreases due to the production of acids and also affects the solubility of CO_2 and H_2 in the medium.

The succinic acid production has been investigated by a set of bacteria such as genetically modified *E.coli*, *Anaerobiospirillum succiniciproducens*, *Mannhemia succiniciproducens* and *Actinobacillus succinogenes*. Among them, *Actinobacillus succinogenes*, a bacterium derived from the bovine rumen, is considered to be a natural succinic acid producer which can produce and accumulate higher amounts of the desired product from glucose as well as from a wide range of sugar-substrates including lactose, xylose, arabinose, fructose and sorbitol [10]. *Actinobacillus succinogenes* is recognised as one of the most suitable microorganisms for the industrial production of succinic acid due to its high tolerance for elevated product concentrations [10].

Until now, there have been very few works studying the dissimilation of glycerol for the succinate production [9]. Most of them have utilised low initial glycerol concentrations due to the decrease of the uptake rate at high glycerol concentrations. The aim of this study is to present an unstructured model that can predict the microbial growth, the substrate and products concentrations during a series of batch fermentations of *Actinobacillus succinogenes* with several initial glycerol concentrations. A factorial design was implemented to determine the optimum environmental conditions of the bioprocess in terms of yield, final succinate concentration and productivity. The necessary kinetic parameters for the model are estimated by fitting the model predictions to experimental observations.

2) Materials and Methods

2.1) Culture strain and inoculum preparation

Actinobacillus succinogenes (ATCC 55617) was obtained from the American type culture collection and it was preserved in cryopreservation vials in -70°C. Preculture of the strain was performed in 100ml Duran bottles containing 50 ml of trypticasein soy broth at 30°C on a rotary shaker at 100 rpm for 1-2 days. Cultivation of the strain was continued using small anaerobic reactors (SAR), each containing 48ml of semi-defined medium [10].

Glycerol was used instead of glucose, as the carbon source in a medium containing per litre: glycerol 10g, yeast extract 5g, $NaH_2PO_4 \cdot H_2O$, 1.16 g; Na_2HPO_4, 0.31 g; NaCl, 1.0 g; $MgCl_2 \cdot 6H_2O$, 0.2 g; $CaCl_2 \cdot 2H_2O$, 0.2 g; B12, 1_g; biotin, 20_g; folic acid, 20_g; thiamine, 50_g; riboflavin, 50_g; niacin, 50_g; pantothenate, 50_g; *p*-aminobenzoate, 50_g; lipoic acid, 50_g; B6, 100_g, $MgCO_3$, 10 g, silicone antifoam, 1mL. Glycerol solution and semi defined medium were separately autoclaved for 15 minutes at 121°C. After inoculation from TSB bottles the SARs were placed on a rotary shaker at 100 rpm and incubated at 37°C for 1-2 days. The volume of the inoculum was around 4 %v/v and CO_2 gas was supplied during the fermentation.

2.2) Batch fermentations in Bench-Top Bioreactor

Batch fermentations were carried out in a 1.8 litres bench-top bioreactor containing a working volume of 0.5 litres. The composition of the medium was similar to the one used in the SARs except for small changes due to the specific experimental design. Thus, glycerol, MgCO3 and $NaBH_4$ concentrations were ranged from (0-40 g/l), (10-30 g/l) and (0-20 g/l) while yeast extract concentration was set to 10 g/l. Moreover, the pH was automatically controlled at a range 6.2-7.0 with the addition of HCl and NaOH (10M solutions). Gaseous CO_2 was supplied at a flow rate of 0.4vvm. The agitation speed set to 200rpm. The inoculum size for the batch fermentation was 10% and all batches were duplicated.

2.3) Analytical techniques

Cell growth was determined spectrophotometrically (spectrophotometer UVmini 1240, Shimadzu,Europa, Germany) by measuring the optical density (OD) at a wave length of 660. In order to ensure that none of the $MgCO_3$ remained undissolved in the sample, 0.5mL of the fermentation sample was diluted with 2 mL of 7% (v/v) hydrochloric acid (HCl) to form soluble magnesium chloride and carbon dioxide. The relationship between 1 OD_{660} and dry cell weight (DCW) per litre was found to be 0.626 g-DCW/litre.

Glycerol concentration was analysed by using a GL6 Analyser (Analox Instruments, UK) which measures the enzymatic oxygen consumption rate. Fermentation products such as succinic acid, acetic acid , formic acid , pyruvic acid and propionic acid were measured by High Perfonmance Liquid Chromatographer (Star Varian Chromatography Workstation) with a UV detector (Prostar 330 PDA). A Hi-Plex H 8 μm 300 × 7.7 mm (Polymer Laboratories) column was used to separate the components preceded by a guard column (PL Hi-Plex H Guard column 50 × 7.7 mm, Polymer Laboratories) at a flow rate of 0.6ml/min using a mobile phase of 0.1% v/v trifluoroacetic acid (TFA) at a column temperature 60°C.

2.4) Factorial design description

The aim of the factorial design is to determine the optimal environmental conditions of the bioprocess in terms of product yield, final concentration and productivity. Moreover, it indicates the crucial environmental parameters that affect the bioprocess and the interactions between them [11] .To do so, a 2^3 factorial design was suggested leading to 8 experiments. Four extra experiments were also carried out at the centre point. These parameters are selected to be the CO_2 supply, the pH level and the redox balance. The values and range of these controlling parameters were selected according to preliminary experiments and previous studies and are shown in Table 1.

Table1: Levels and range of the factorial design variables

Variables	Range and Levels		
	-1	0	+1

$MgCO_3$ (g/l)	X_1	10	20	30
pH	X_2	6.3	6.6	6.9
$NaBH_4$ (g/l)	X_3	0	10	20

The controlling parameters were coded according to the following equation:

$$X_i = \frac{X_i^r - X_i^{ro}}{\Delta X} \quad (1)$$

where X_i is the coded value of the variables, X_i^r and X_i^{ro} are the real values of the variables in level -1, +1 and 0 respectively and ΔX is the step change.
The following linear polynomial was used for the estimation of the dependent variable.

$$Y_i = \beta_0 + \sum_{i=1}^{3} \beta_i X_i + \sum_{i=1}^{3} \beta ij X_i X_j + \beta_{ijk} X_i X_j X_k \quad (2)$$

where Y_i is the estimated variable, β_o is the intercept of the polynomial, β_i is the linear coefficients, and β_ij and β_{ijk} are the interaction coefficients.
The adequacy of eq.(2) was checked by the Fisher criterion with level of confidence 5% while the importance of the coefficients was checked by the student-t distribution also with a 5% level of confidence[2].

2.5) Model studies

A modified Monod model was used to describe the growth kinetics considering both substrate and product inhibition.

In preliminary experiments, it was estimated that there is an excessive substrate inhibition on microbial growth. Therefore, the following modified Monod model was used to describe the substrate inhibition [12],[13],[14]:

$$\mu = \mu_{\max} \cdot \frac{S}{S + K_S + (S^2/K_I)} \quad (3)$$

where μ is the specific growth rate (h^{-1}), μ_{max} is the maximum specific growth rate (h^{-1}), S is the substrate concentration (g/L), K_s is the substrate saturation constant (g/L) and K_I is the substrate inhibition constant (g/L).

Furthermore, during the fermentation the formation of products can act as an inhibition factor and can gradually affect the cell growth. Thus, in addition to eq.(1) a product inhibition term was introduced [13].

$$\mu = \mu_{\max} \cdot (1 - \frac{P_i}{P_i^*})^n \quad (4)$$

where P is the concentration of the product during the fermentation (g/L), P^* is the succinic acid concentration above which cells do not grow (g/L), *n* is the empirical constant which describes the linearity or not of the system (dimensionless) and the index indicates the number of by-products.

Combining eq.(3) and eq.(4), a final extended expression is developed (eq. 5) which takes into consideration substrate inhibition and multi inhibition effects of products.

$$\mu = \mu_{\max} \cdot (\frac{S}{S + K_S + (S^2/K_I)})(1 - \frac{P_i}{P_i^*})^n \quad (5)$$

The cell growth rate (dX/dt) can then be described by equation (6).

$$\frac{dX}{dt} = \mu \cdot X \tag{6}$$

where X is the cell concentration (g/l)

Finally, the substrate consumption rate (dS/dt) and the products formation rate (dP/dt) can be described by Luedking-Piret model (eq. 7 and 8 respectively).

$$\frac{dS}{dt} = -\alpha \frac{dX}{dt} - \beta X \tag{7}$$

$$\frac{dP}{dt} = \gamma \frac{dX}{dt} + \delta X \tag{8}$$

where α is the constant term for substrate consumption (g-S/g-X), β is the maintenance coefficient of the substrate (g-S/g-X h), γ is the constant growth term of product formation (g-P/g-X) and δ is the maintenance coefficient of the products (g-P/g-P h).

2.6) Parameter Estimation

The eight parameters of the model proposed were estimated using nonlinear, weighted, least squares method by minimizing the sum of squared errors between the estimated and experimental values. Simulated annealing (SA), a stochastic optimization algorithm implemented in MATLAB was used. Simulated annealing can avoid local minima and probabilistically converges to the region around the global minimum. A deterministic optimisation step using Successive Quadratic Programming was implemented next to pinpoint precisely the global minimum. In order to avoid potential unrealistic optima all parameters were constrained within limits found in the literature for similar systems [2], [12], [13], [14]. These bounds are presented together with the values obtained from the combined optimization algorithm in Table 2. The initial values of the ODEs were calculated according to the initial conditions of each experiment.

3) Results and Discussion

Preliminary experiments on glycerol for succinic acid production by *Actinobacillus succinogenes* showed that this substrate, compared to glucose, is not very amenable to biodegradation and results in low productivities and final succinic acid concentrations due to possible substrate inhibition. In this study we have developed a priori a factorial design experiment to examine the effect of the most crucial parameters.

Factorial design results indicated the significance of the most important parameters (CO_2 supply, pH value and redox balance) and their interactions. Succinic acid was affected by all three parameters and its production was amended due to $MgCO_3$ and $NaBH_4$ addition at low pH values. According to Student's t-test $MgCO_3$ had the strongest positive linear effect on the predicted values followed by the pH values and $NaBH_4$. Linear and interaction coefficients were calculated from the obtained experimental values and optimum conditions were estimated by optimising the predicted value (eq 2) in terms of productivity, final succinic acid concentration and yield. Optimum conditions were assessed to be at 40 g/L $MgCO_3$, 25 g/L $NaBH_4$ and 6.4 pH value. The experiments that were carried out at this point to validate the results illustrated good agreement with the predicted values (data not shown). Furthermore, the interactions between the controlling parameters were estimated. It seems that there is a strong interaction between both $NaBH_4$ and $MgCO_3$ and the pH value. Significant interactions were also noted between $NaBH_4$ and $MgCO_3$.

At the optimum conditions, batch fermentations with initial glycerol concentration of 0, 5, 10, 15, 20, 30, 40, 50 g/L were carried out and the unstructured model described above was used to simulate the experimental data. The proposed model includes both substrate and product inhibition. Simulation results using model for cell growth, substrate concentration and succinic acid formation are shown in figure 1. Since in almost all the experiments the by-product formation was very low compared with succinic acid; acetic acid, formic acid, ethanol and pyruvic acid product inhibitions were not included in the model.

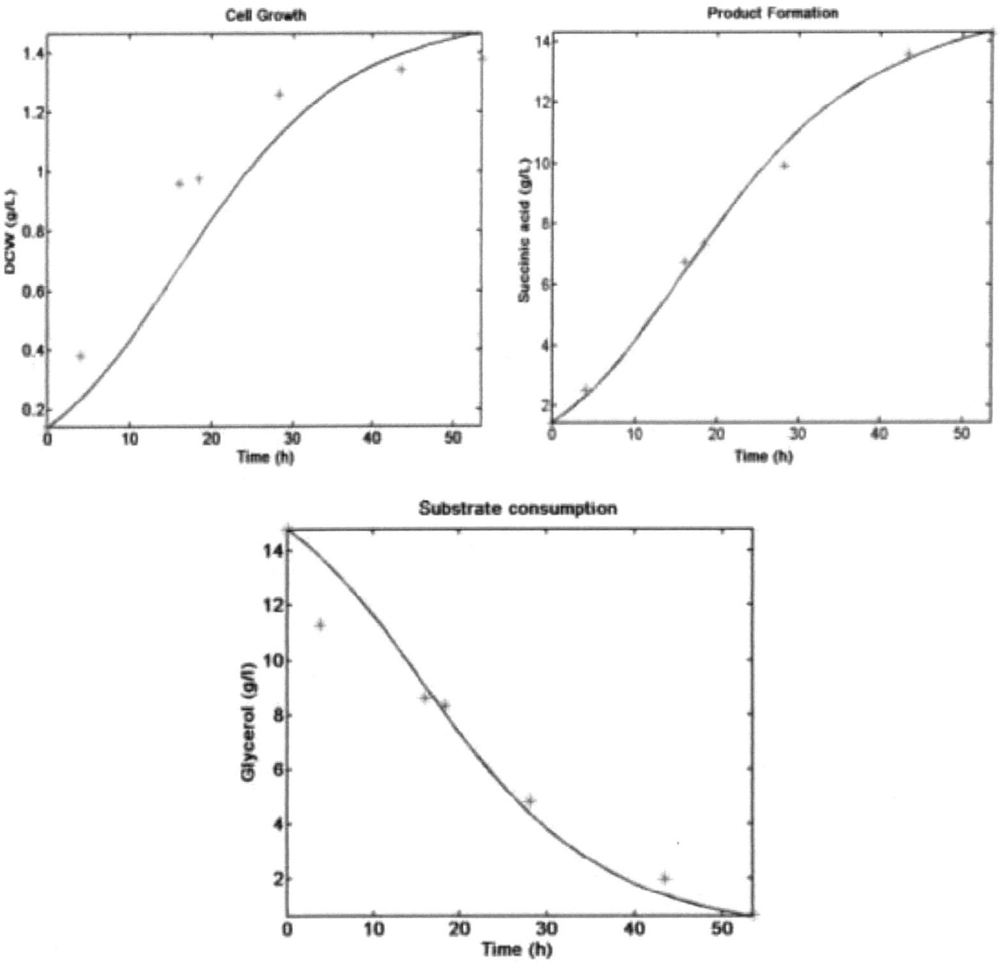

Figure 1: Experimental and simulation data of glycerol, biomass and succinic acid

The critical concentration of succinic acid P* above which cells do not grow was measured in a previous study [12] and it was found to be equal to 155 g/L; this value indicates the major tolerance of *Actinobacillus succinogenes* to high succinic acid concentrations. The parameters of the model were estimated using a combination of stochastic and deterministic optimisation and the obtained results are shown in table 2. The model can adequately predict the kinetics of the fermentation for a wide range of initial glycerol concentrations. Although, it is the first time a model has been developed for this system, the obtained parameter values were compared with results from similar studies such as 1,3-PD production [3], and succinic acid production from glucose [12], [13] and showed good agreement. Nevertheless, it should be noted that parameter values of such an unstructured model have limited physical significance due to the fact that they are based on an overall reaction for all the phenomena involved in the conversion of substrate to biomass.

Table 2: Parameters of the model

Parameters	Units	Values	Description	Bounds
μ_{max}	h^{-1}	1.39	Maximum specific growth rate	0.07-1.6
K_s	g/L	2.5	Substrate saturation constant	0.04-4
K_I	g/L	80	Substrate inhibition constant	1-80
n	-	0.5	empirical constant	0.4-1.4
α	g-S/g-X	4.67	Constant term for substrate	0.6-5.4
β	g-S/g-X h	0.001	Substrate maintenance coefficient	0.001-1.2
γ	g-P/g-X	2.02	Constant growth term of product	0.4-2.8
δ	g-P/g-X h	0.018	Product maintenance coefficient	0.001-1.2

The maximum product yield, final concentration and productivity were found to be 0.75mol-S.A/mol-Gly, 28 g/L S.A. and 0.4 g-S.A./L h respectively on a medium containing initially 29.3 g/L glycerol. No cell growth was observed above 40g/l initial glycerol concentration. However, these results are significantly lower than the corresponding ones for 1,3-PD production from glycerol [3] and for succinic acid production from glucose [12]. They are, however, the highest observed for this system and further experimental investigations to obtain better results are needed.

Conclusions

In this study, an empirical model was developed to describe *Actinobacillus succinogenes* growth on glycerol for succinic acid production. The modified Monod expression model includes substrate and product inhibition and can adequately predict batch fermentation production rates for a wide range of initial glycerol concentrations. Experiments were carried out at optimum environmental conditions indicated by a 2^3 factorial design. The parameter values obtained were in good agreement with values in similar studies. Although, succinic acid production from glycerol is less favourable compared to the formation of 1,3-propanediol from glycerine and succinic acid production from glucose, this process seems to be a promising alternative increasing the sustainability of biodiesel production. Further investigation is needed to optimise the bioprocess. Fed-batch fermentations and recycling biomass techniques may improve the process significantly. Also, more detailed models that can give insights into the intracellular reactions and phenomena are needed. These models can be used for optimising the succinic acid metabolic pathways simultaneously minimising by-product streams.

References

[1] J.C.Thompson, B.B.He, Characterization of crude glycerol from biodiesel production from multiplefeedstocks, Appl. Eng. Agric, 22, (2006), 261-265

[2] Zheng, ZM; Hu, QI; Hao, H. Statistical optimization of culture conditions for 1,3-propanediol by *Klebsiella pneumoniae* AC 15 via central composite design, Bioresour. Technol., 99, (2008), 1052-1056

[3] Zeng, AP; Ross, A; Biebl, H., Multiple product inhibition and growth modelling of *Clostridium butyricum* and *Klebsiella pneumonia* in glycerol fermentation, Biotechnol.Bioeng., 44, (1994), 902-911

[4] Dharmadi, Y; Murarka, A; Gonzalez, R, Anaerobic fermentation of glycerol by *Escherichia coli*: A new platform for metabolic engineering Biotechonol. Bioeng., 94, (2006), 821-829

[5] McKinlay, JB; Vieille, C; Zeikus, JG, Prospects for a bio-based succinate industry, Appl Microbiol Biotechnol., 76, (2007), 727-740

[6] Van der Werf, MJ; Guettler, MV; Jain, MK., Environmental and physiological factors affecting the succinate product ratio during carbohydrate fermentation by *Actinobacillus* sp. 130Z, Arch Microbiol, 167, (1997) 332-342

[7] McKinlay, James B; Vieille, Claire, 13C-metabolic flux analysis of *Actinobacillus succinogenes* fermentative metabolism at different NaHCO3 and H2 concentrations. Metab Eng, 10, (2008) 55-68

[8] Samuelov, NS; Lamed, R; Lowe, S, Zeikus J.G. Influence of CO2-HCO3- Levels and pH on growth succinate production and enzyme activities of *Anaerobiospirillum succiniciproducens*, Appl Env Microbiol., 57, (1991) 3013-3019

[9] Lee, PC; Lee, WG; Lee, SY, et al., Succinic acid production with reduced by-product formation in the fermentation of *Anaerobiospirillum succiniciproducens* using glycerol as a carbon source, Biotechnol. Bioeng., 72, (2001) 41-48

[10] Guettler, MV; Rumler, D; Jain, MK, *Actinobacillus succinogenes* sp. nov., a novel succinic-acid-producing strain from the bovine rumen, Intern. J. System. Bacteriol., 49, (1999) 207-216

[11] Cochran, W.C. and Cox, Experimental Designs, (1957) New York: J. Wiley & Sons Inc

[12] Sze Ki C. Lin, C. Du, A. Koutinas, R. Wang, C. Webb, Substrate and product inhibition kinetics in succinic acid production by *Actinobacillus succinogenes* Biochem. Eng. J., 41 (2008) 128-135

[13] Song, H; Jang, SH; Park, JM, et al, Modeling of batch fermentation kinetics for succinic acid production by *Mannheimia succiniciproducens*, .Biochem. Eng. J. 40 (2008) 107-115

[14] J F. Andrews, A Mathematical Model for the Continuous Culture of Microorganisms Utilizing Inhibitory Substrates, Biotechonol Bioeng, 10 (1968) 707-723

U.S. Department of Energy

Impacts of New Technologies and Policies on Biofuels Production and Trade

Bhima Sastri, Ph.D.

Audrey Lee, Ph.D.

Office of Policy and International Affairs,
Department of Energy, Washington DC

Thomas Alfstad

Brookhaven National Laboratory

Richard Bain

National Renewable Energy Laboratory

November 16-21, 2008

AIChE Annual Meeting

Outline

- Policies - U.S. and worldwide
- Model Methodology
- Technologies for Biofuels
- Results
  - Reference Case
  - Scenario Cases
    - CO_2 price, oil price, E20
- Conclusions

Energy Independence & Security Act of 2007

- Renewable fuel standards for feedstocks & GHG emissions:
  - Renewable Fuel: Fuel derived from renewable biomass (Including corn starch)
  - Advanced Biofuel: Renewable fuel (not from corn starch) with fewer GHG emissions
  - Cellulosic Biofuel: Advanced biofuel from cellulose, hemicellulose or lignin
  - Biomass-based Diesel: Advanced biofuel replacing diesel
- Requirements are nested:
  - Firm requirements for cellulosic biofuels and bio-diesel.
  - Advanced biofuels may be all cellulosic and bio-diesel.
  - Renewable fuels may be all advanced biofuels.
- Waivers available – financial buyout for cellulosic biofuels.

Energy Independence & Security Act of 2007

Renewable Fuels Standard

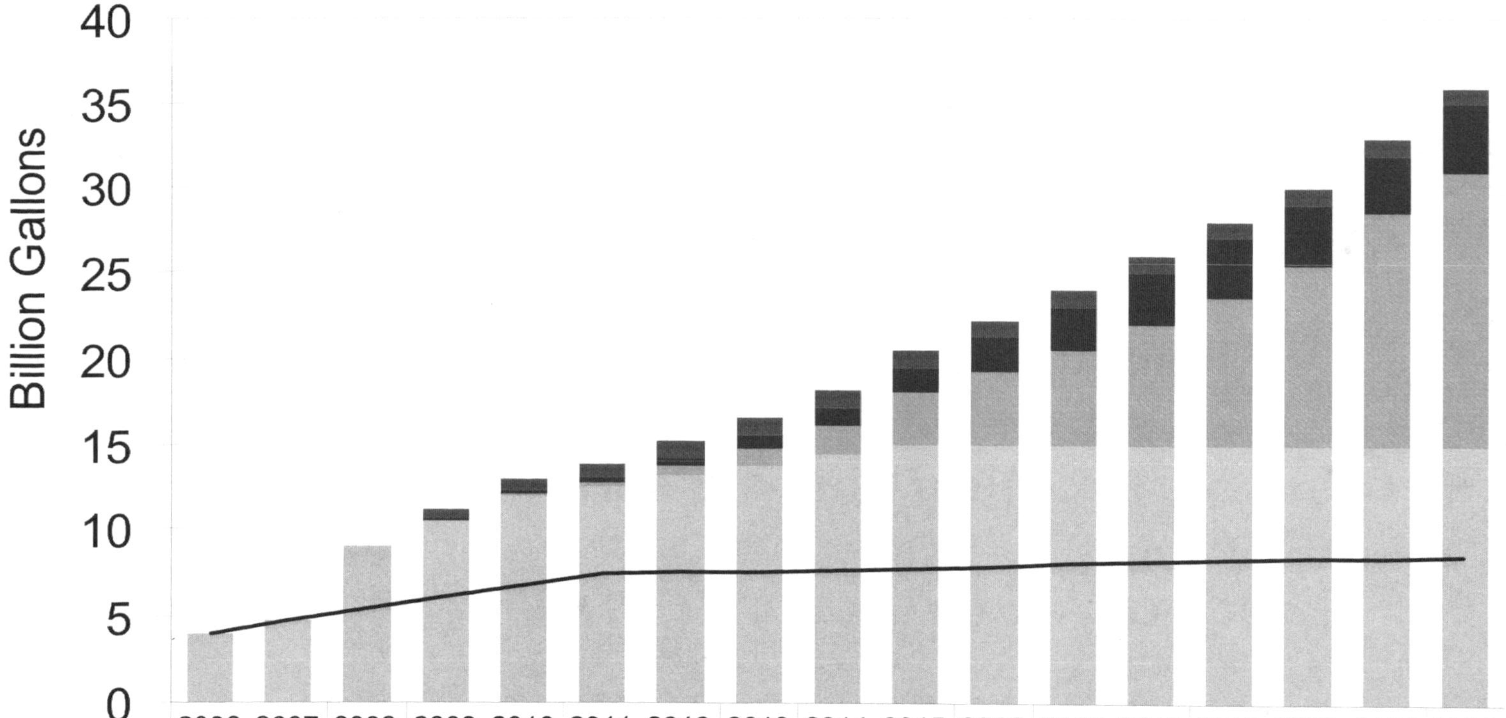

	2006	2007	2008	2009	2010	2011	2012	2013	2014	2015	2016	2017	2018	2019	2020	2021	2022
Biomass based Diesel				0.5	0.65	0.8	1	1	1	1	1	1	1	1	1	1	1
Any Advanced	0	0	0	0.1	0.2	0.3	0.5	0.75	1	1.5	2	2.5	3	3.5	3.5	3.5	4
Cellulosic Advanced					0.1	0.25	0.5	1	1.75	3	4.25	5.5	7	8.5	10.5	13.5	16
Any Biofuels	4	4.7	9	10.5	12	12.5	13.2	13.8	14.4	15	15	15	15	15	15	15	15
Old RFS	4	4.7	5.4	6.1	6.8	7.4	7.5	7.6	7.7	7.8	7.9	8.1	8.2	8.3	8.4	8.5	8.6

EISA'07 RFS Restrictions

- **Minimum GHG Reductions:**
  - Renewable Fuel: 20%
  - Advanced Biofuel: 50%
  - Cellulosic Biofuel: 60%
  - Biomass-Based Diesel: 50%
- **Land Use Must Be:**
  - Cleared or under cultivation & non forested prior to EISA'07 (crops)
  - Managed plantations (trees)
- **Feedstocks May Include:**
  - Crops from previously cleared, non-forested land
  - Biomass from private forest lands*
  - Algae
  - Separated yard and food wastes
- **Feedstocks Do Not Include:**
  - Biomass from ecologically sensitive, protected lands
  - Biomass from federal forest lands

*Includes native-American lands, privately held forests and tree plantations

Worldwide National Policies

Country/ region	Gasoline tax	2010 Biofuel tax exemption	Ethanol tariffs	Other, modeled	Other, not-modeled in current study
Australia	$1.40/gal	100%	90¢/gal		
Canada	$0.25/gal	100%	20¢/gal		5% market share by 2010
China	$0.15/gal	100%	0		15% market share 2015
Central & S. America	$0.70/gal	50%	27¢/gal	Subsidy for hydrous ethanol & FFV; Brazil blending requirement of 20-25%	
Europe	$2.80/gal	90%	90¢/gal	5.5% market share 2010 10% market share 2020	
India	$1.90/gal	0%	200%		5% market share by 2015
Japan	$1.85/gal	90%	17%	500 million liters gasoline equivalent by 2010	
S. Korea	$3.02/gal	90%	0		
USA	$0.42/gal	51¢/gal	54¢/gal	36 billion gallons renewable fuels 2022	

MARKAL Model Structure

Energy Technology Perspectives Model

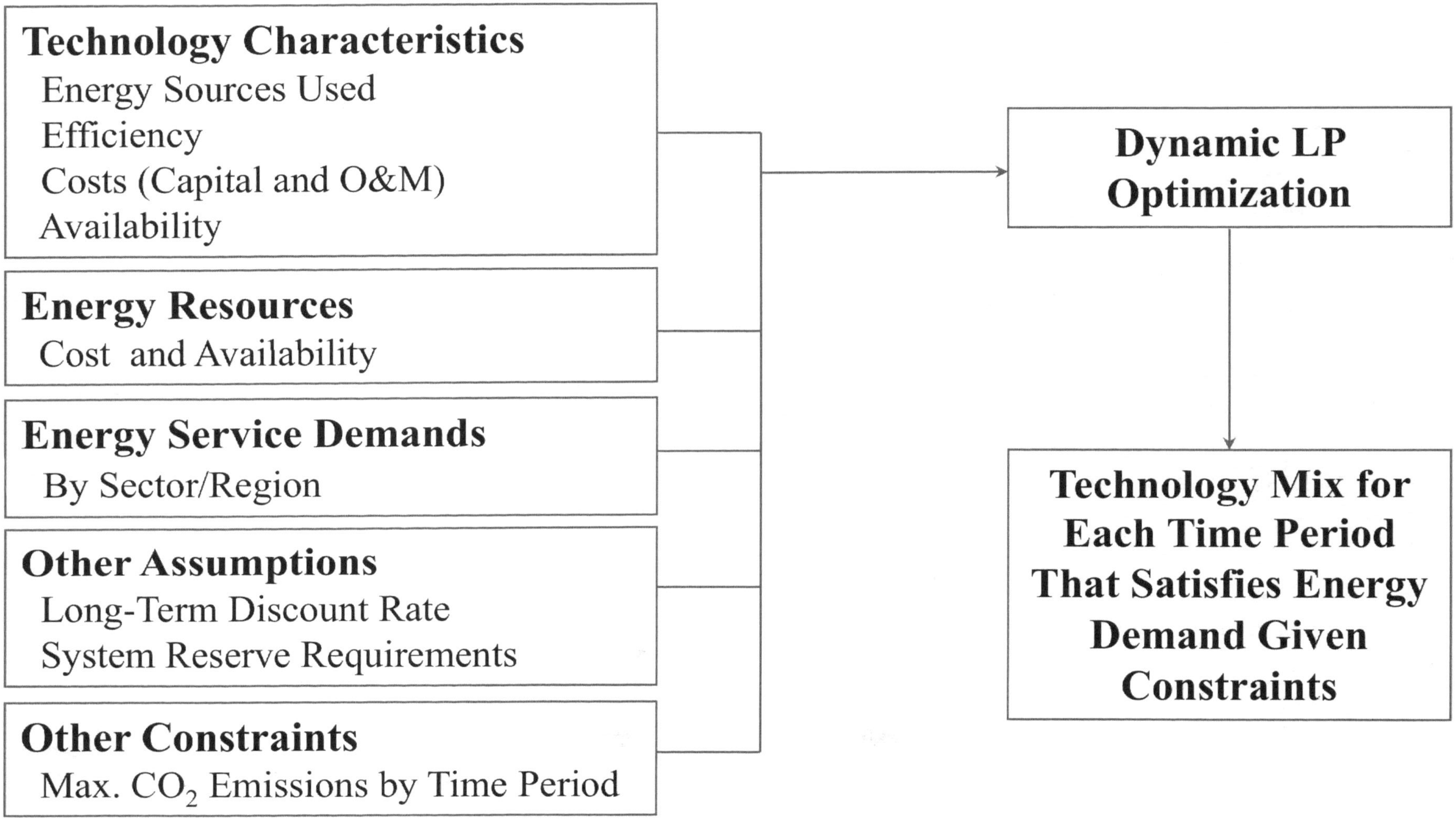

Updates to ETP Model-Feedstocks

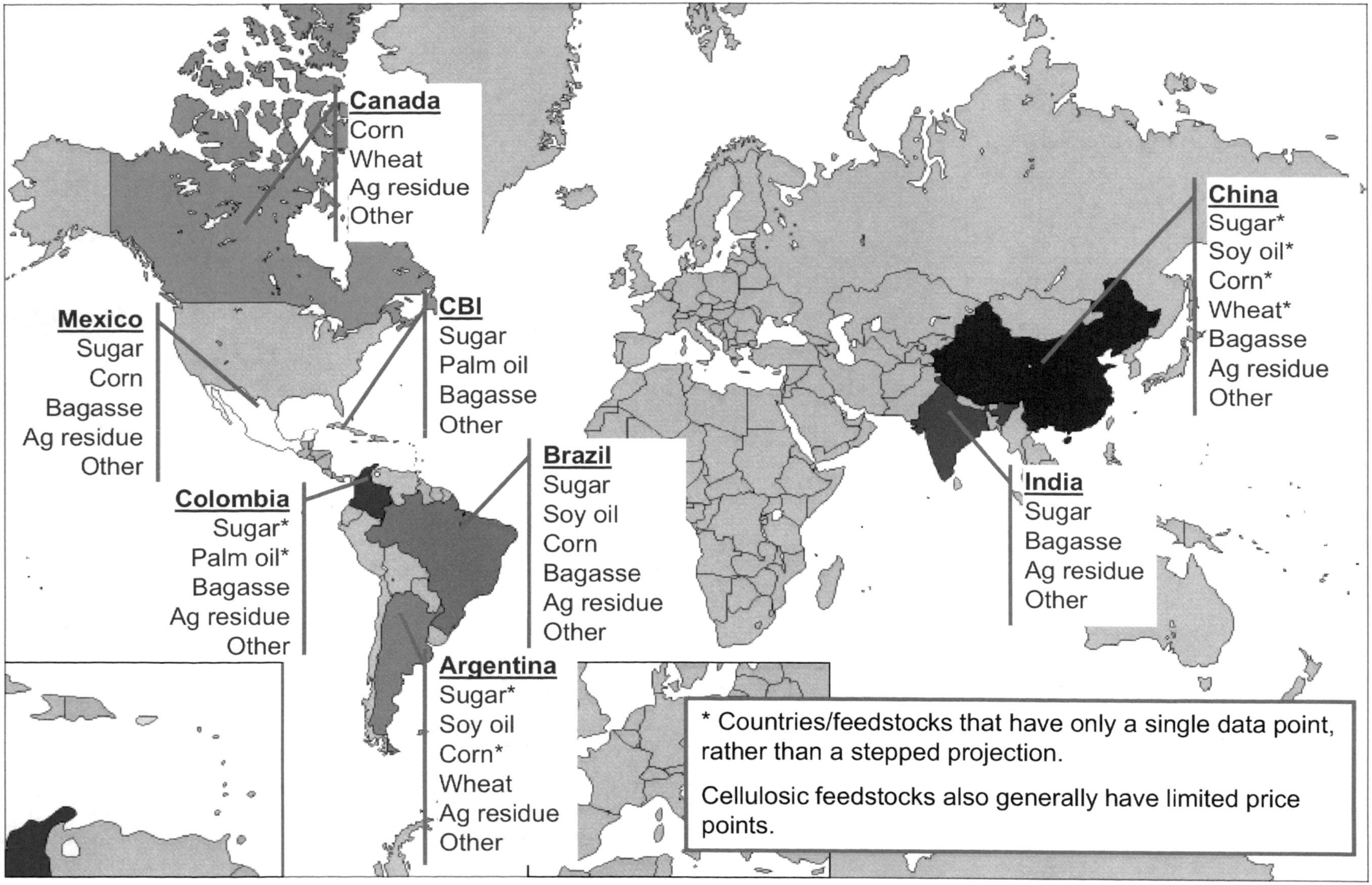

Updates to ETP Model-Technologies

Feed stock	Source	Conversion Technology	Product	Distribution/ Consumption
Sugar	Sugarcane	Sugar-ethanol mill	Ethanol	• New distribution infrastructure required • Consumption limited to E10 for most of existing vehicle fleet • Higher blends (i.e. E85) can be used in small portion of fleet
Starch	Corn Wheat	Dry mill	Ethanol	
Cellulose	Bagasse/other agricultural residues Forestry residues Energy crops	Biochemical conversion	Ethanol	
		Thermo-chemical alcohol synthesis	Ethanol/ higher alcohols	
		Fischer-Tropsch synthesis	Distillates, naphtha	• Products are refining feedstocks • Compatible with conventional fuel infrastructure
Oil	Oil Palm Soybean	Transesterification	Biodiesel (FAME)	• Can be blended with petrodiesel at high ratios in most applications

Conversion Technologies

- Ethanol
  - Sugarcane
  - Dry Mill – Corn, Wheat
  - Thermo-chemical Process for Cellulosic Feedstocks (Alcohol Synthesis)
  - Biochemical Process for Cellulosic Feedstock
- Biodiesel
  - Soy Oil
  - Palm Oil
- Biomass-to-Liquids products
  - Thermo-chemical Process for Cellulosic Feedstocks (Fischer-Tropsch)

Dry Corn Mill

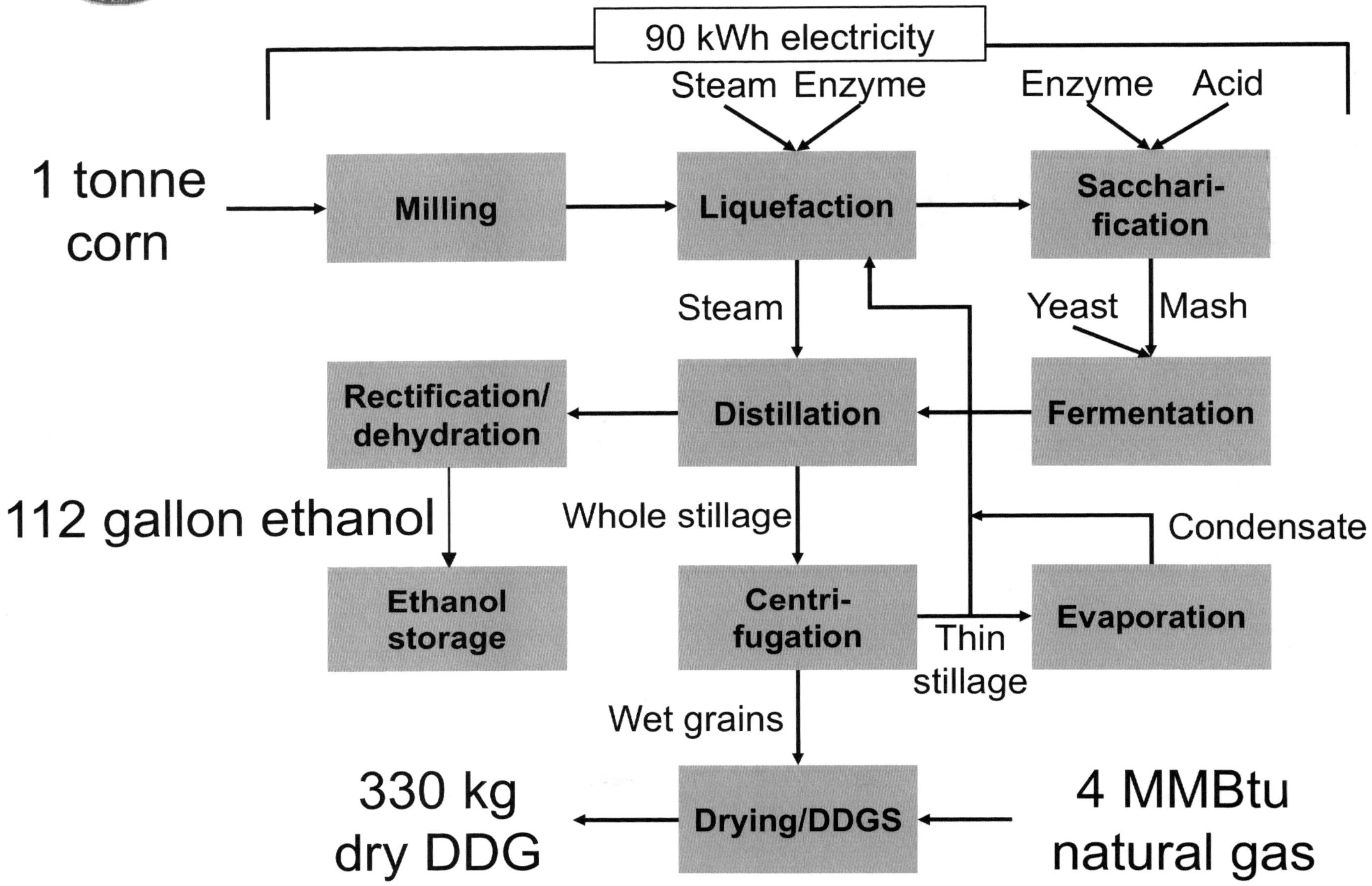

Sugarcane Mill

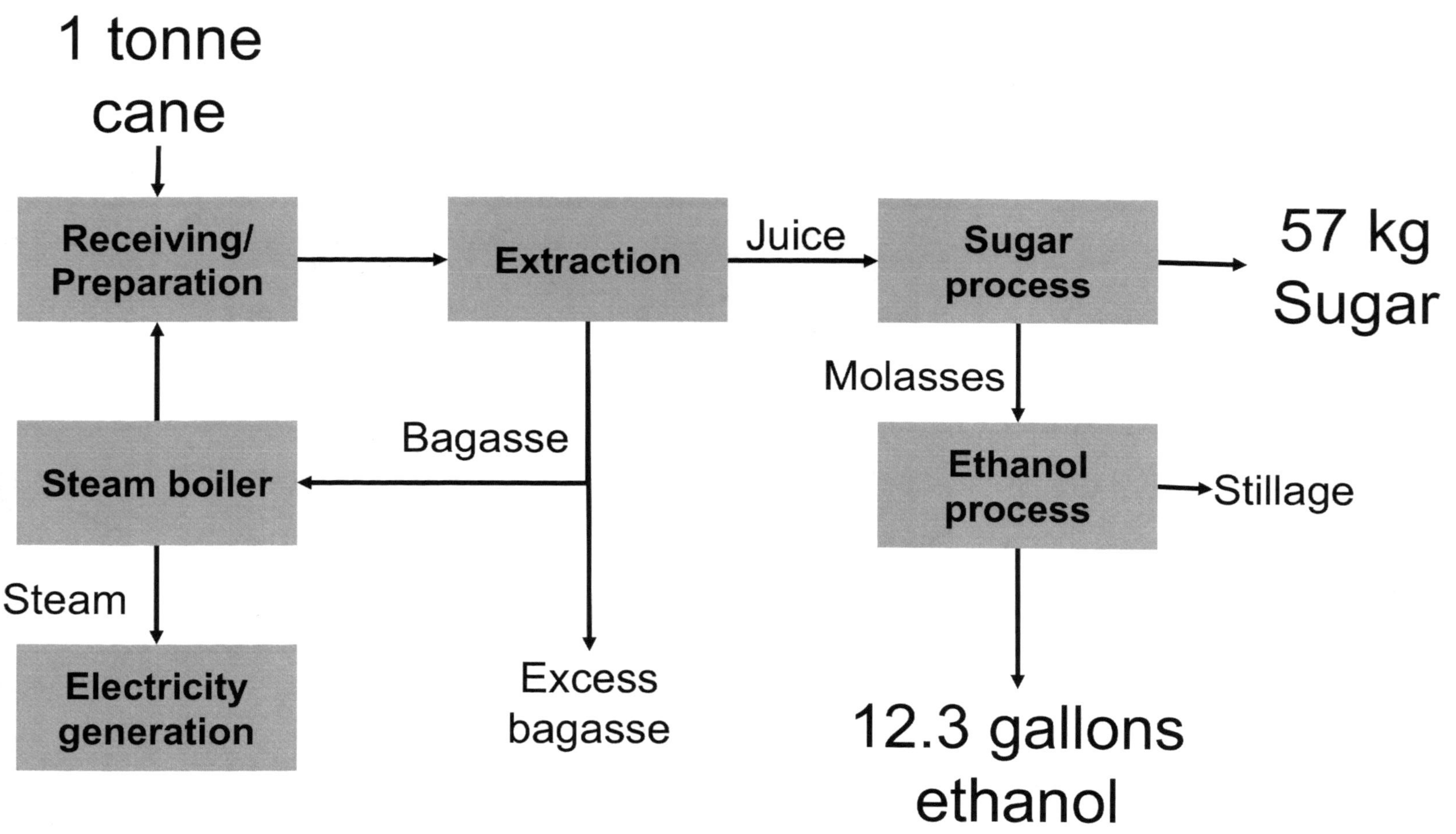

Bio-chemical Conversion

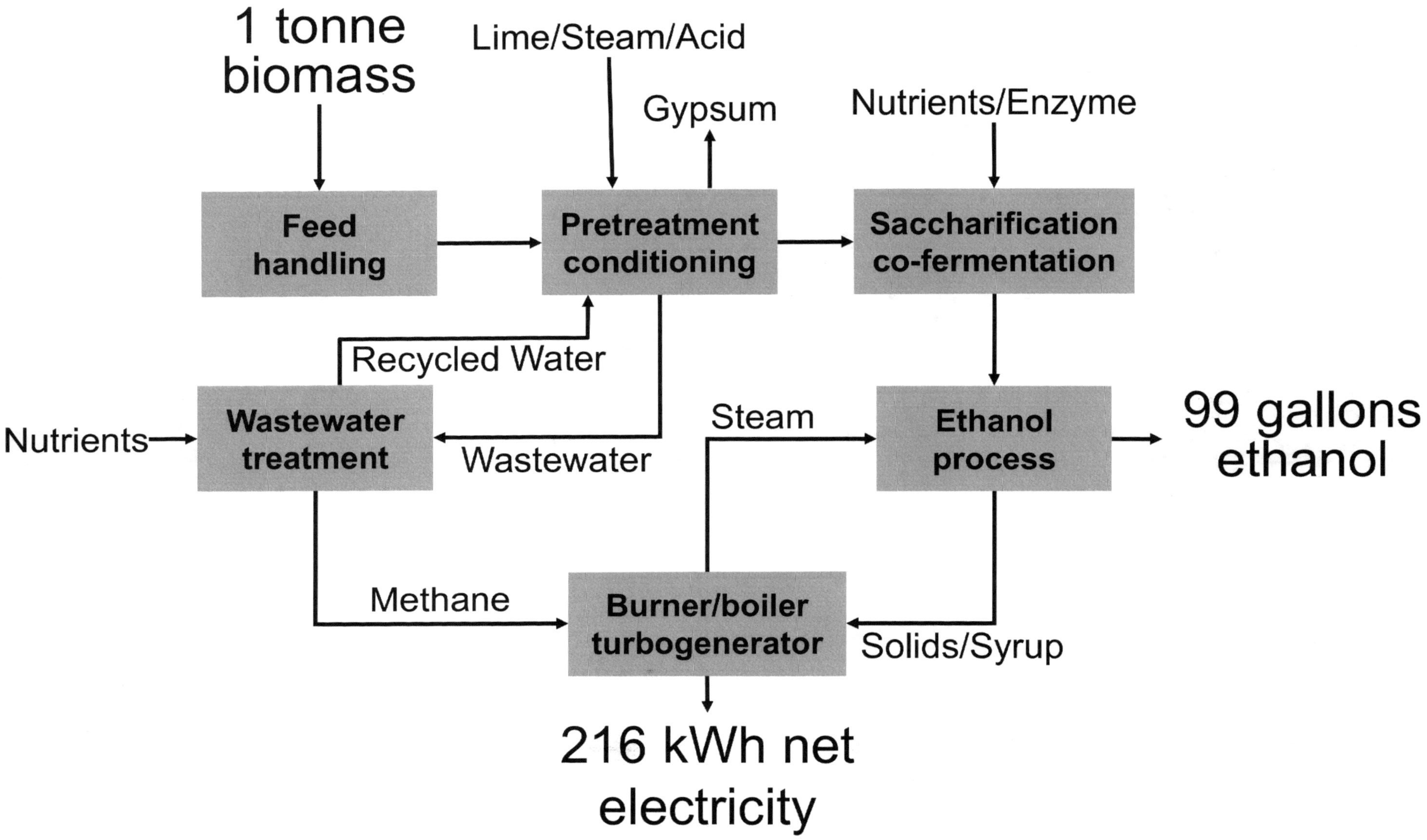

Thermo-chemical Conversion

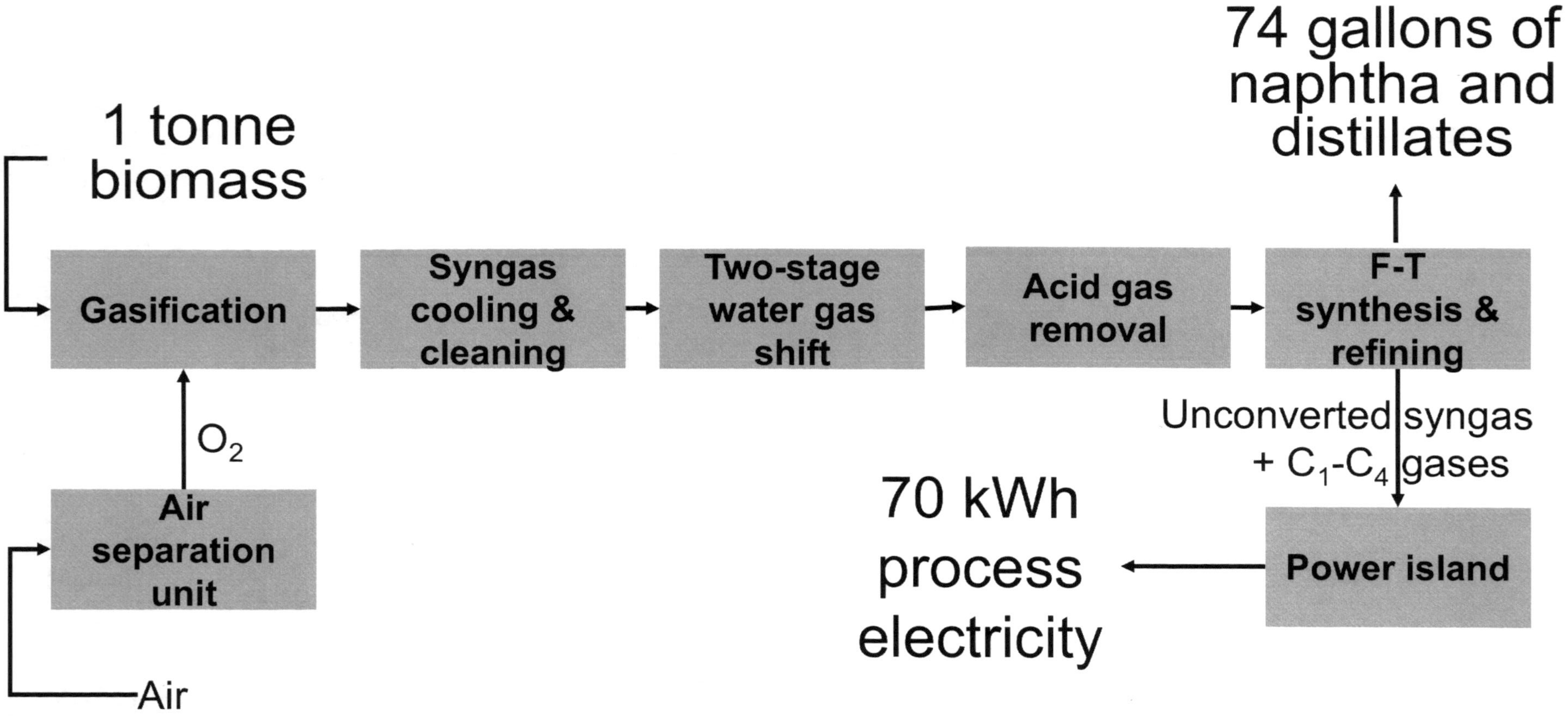

Thermo-chemical Conversion

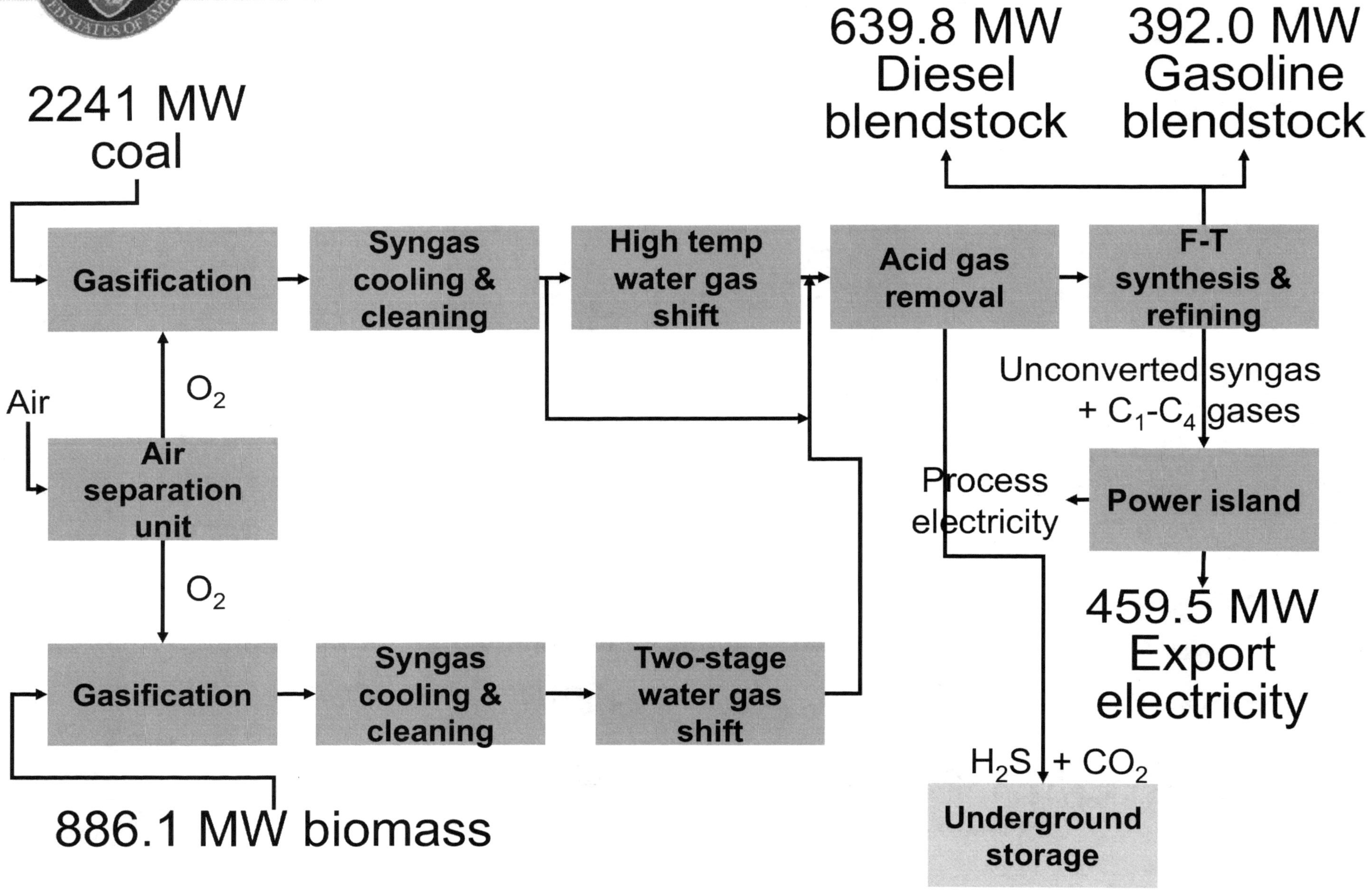

Reference Case Assumptions

- EISA Renewable Fuel Standard
- $1.01/gallon cellulosic biofuel subsidy extended until cost competitive
- $1.00/gallon biodiesel subsidy
- Blenders' ethanol credit of $0.51/gallon and Tariff of $0.54/gallon expire in 2010
- Includes existing national biofuels policies worldwide

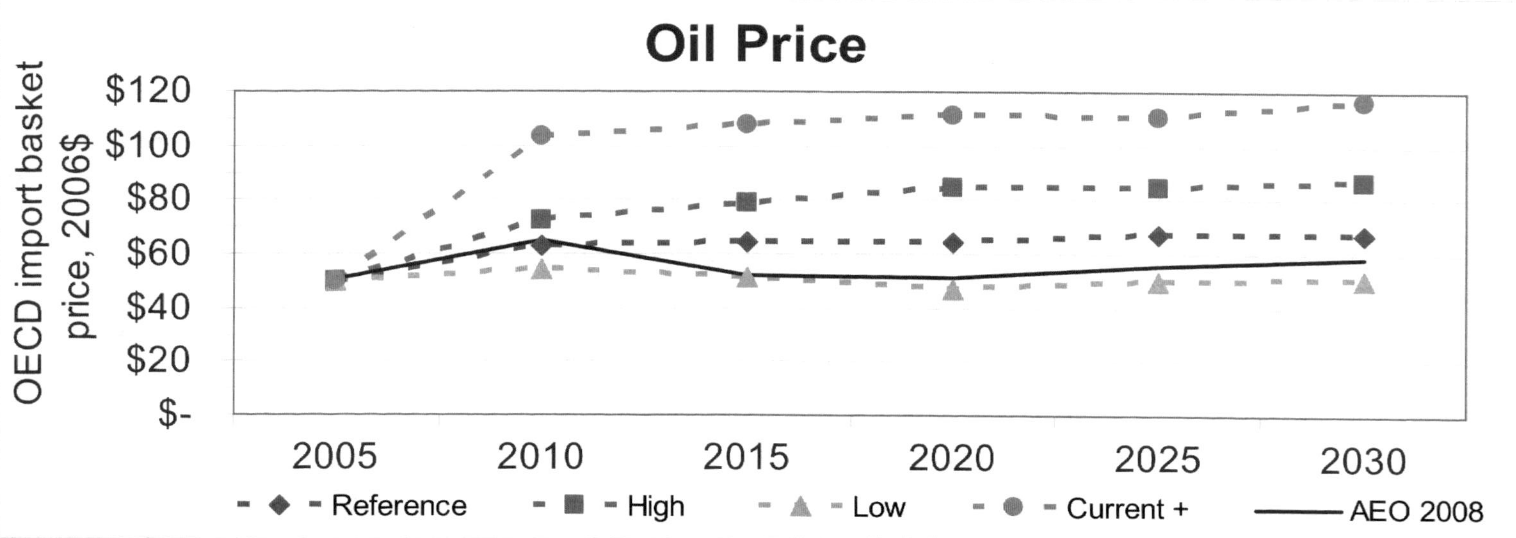

Oil prices are OECD import basket prices (typically much lower than NYMEX oil prices).

Worldwide Biofuels Production

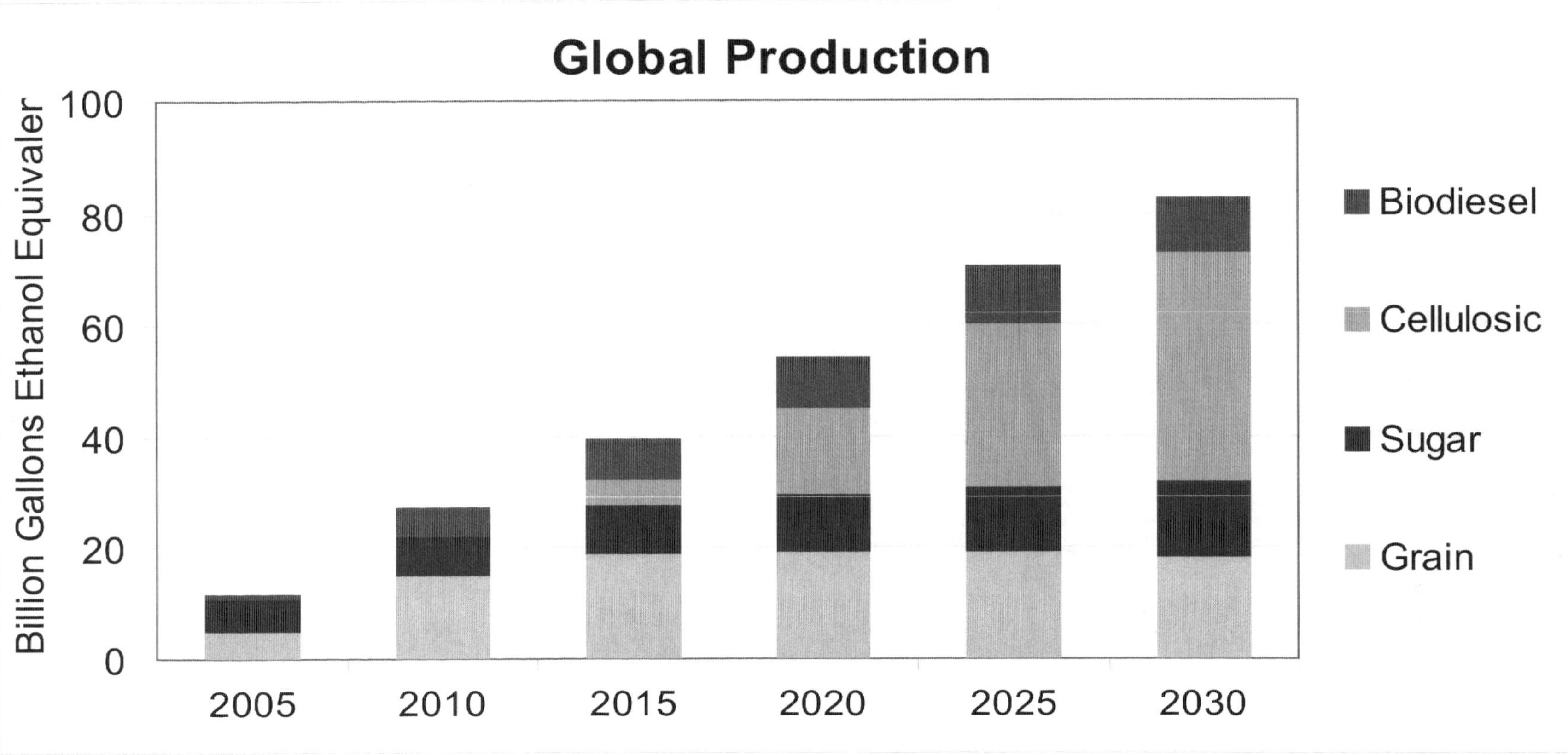

- **Grain production levels off after 2015**
- **Large growth in cellulosic biofuels**
- **Subsidy for early cellulosic plants is crucial to this growth**

Reference Scenario vs. AEO

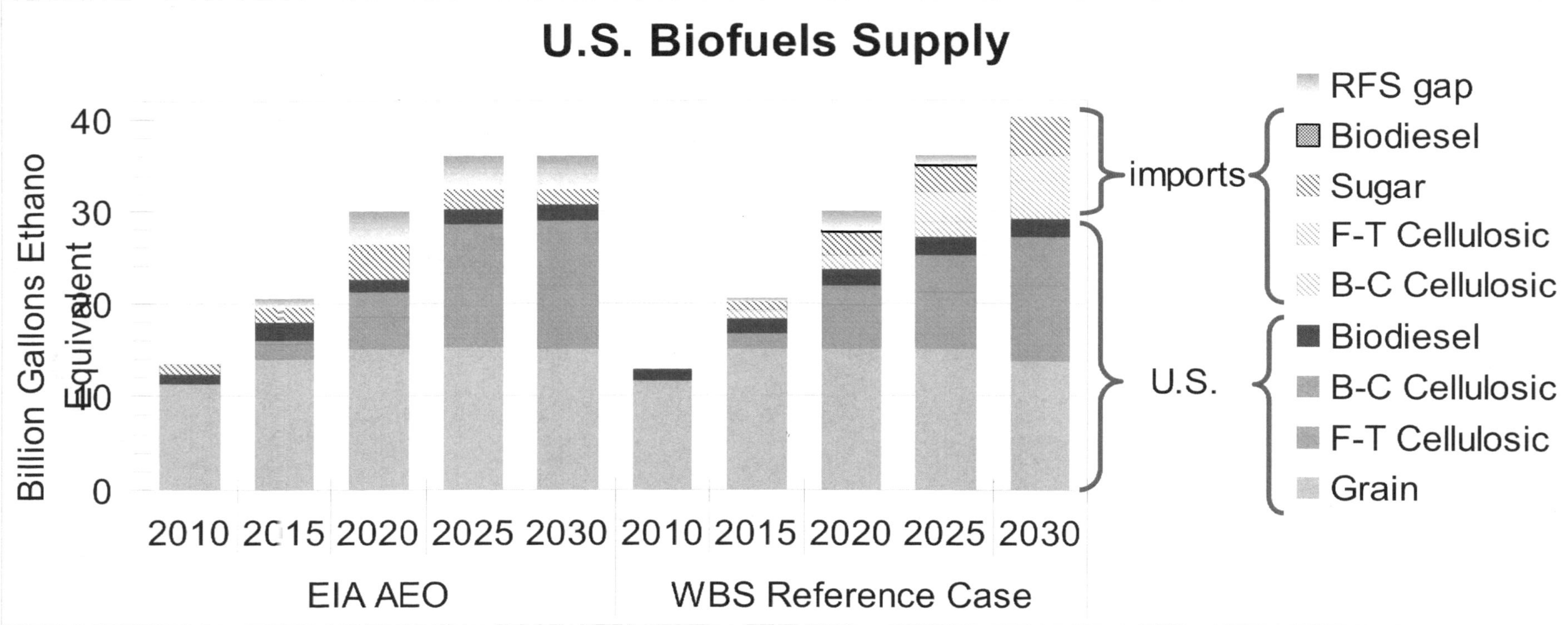

- We project more imports than EIA's Annual Energy Outlook.
- Both domestic & imported cellulosic biofuels will contribute to meeting the mandate.
- Main challenge is building cellulosic plants fast enough.

Scenarios Modeled

Policy Scenarios

Tariff/Credit Extension
Credit Extension
$50/tCO_2$ (global)
E20 Certification
Grower's payment

Market Scenarios

High/Low Feedstock Supply
Low/High/Higher Oil Price
Higher share of Brazilian sugar to ETOH
High Oil Price + High Feed
Low Oil Price + Low Feed

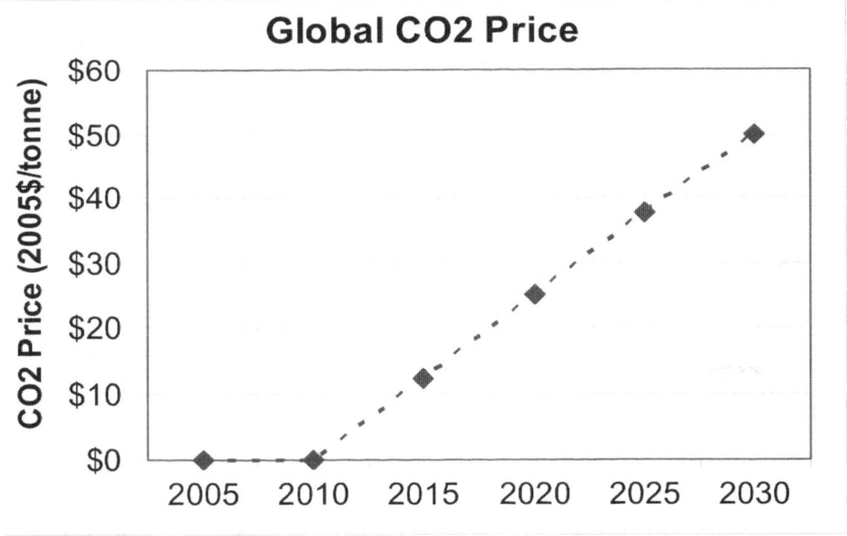

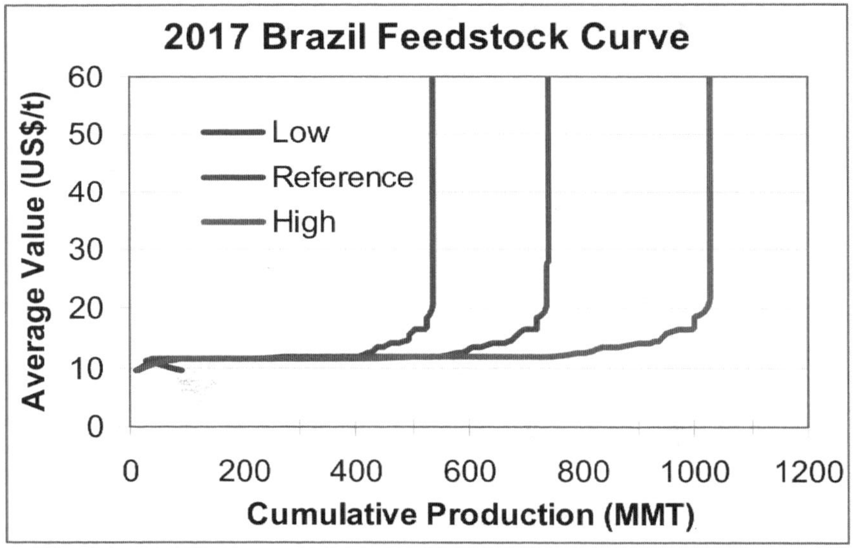

CO_2, Oil Price Scenarios (U.S.)

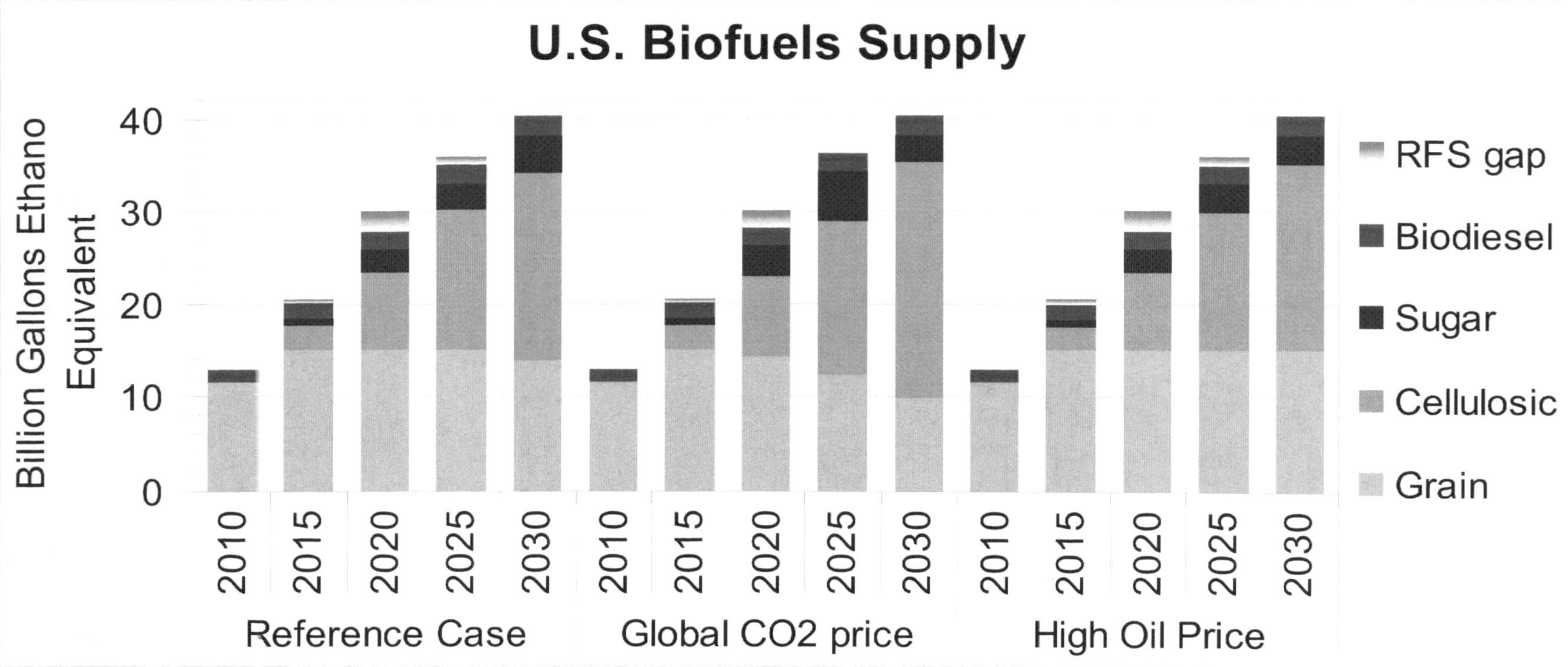

- **Global CO_2 price:**
  - **RFS is met after 2025**
  - **High oil price: little change from reference because buy-out for cellulosic varies with oil price**

CO_2, Oil Price Scenarios (U.S.)

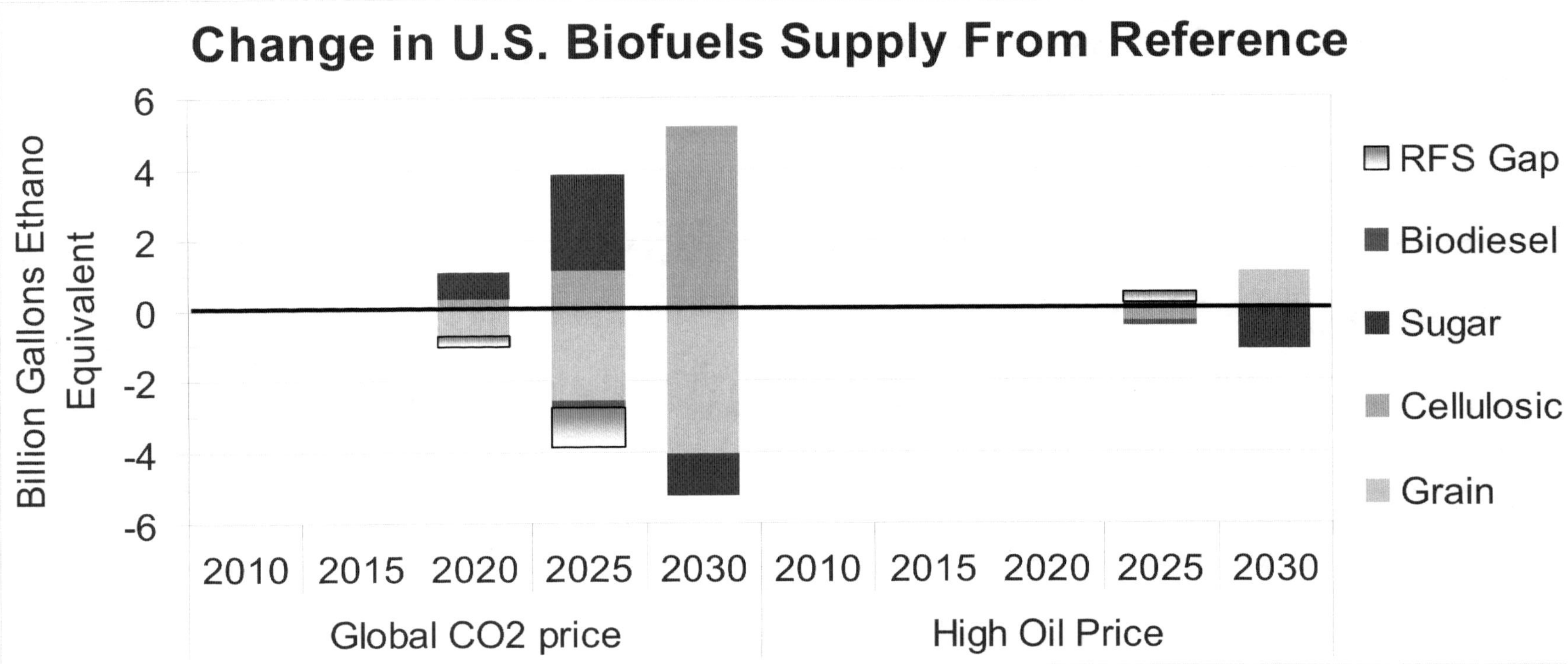

- **Global CO_2 price:**
  - **Closer to meeting RFS than Reference Case**
  - **Sugar replaces corn and fills in RFS gap in 2025**
  - **Cellulosic replaces sugar and corn in 2030**
- **High oil price: slightly more corn in place of sugar**

The barrier to meeting RFS?

Biofuels Supply or Infrastructure

- We used the E20 certification scenario to investigate whether ethanol infrastructure was the barrier to meeting the RFS.
- The E20 scenario is a hypothetical scenario that allows increased use of ethanol without new pipelines, fueling stations, and flex fuel vehicles.

E20 Scenario: U.S. Supply

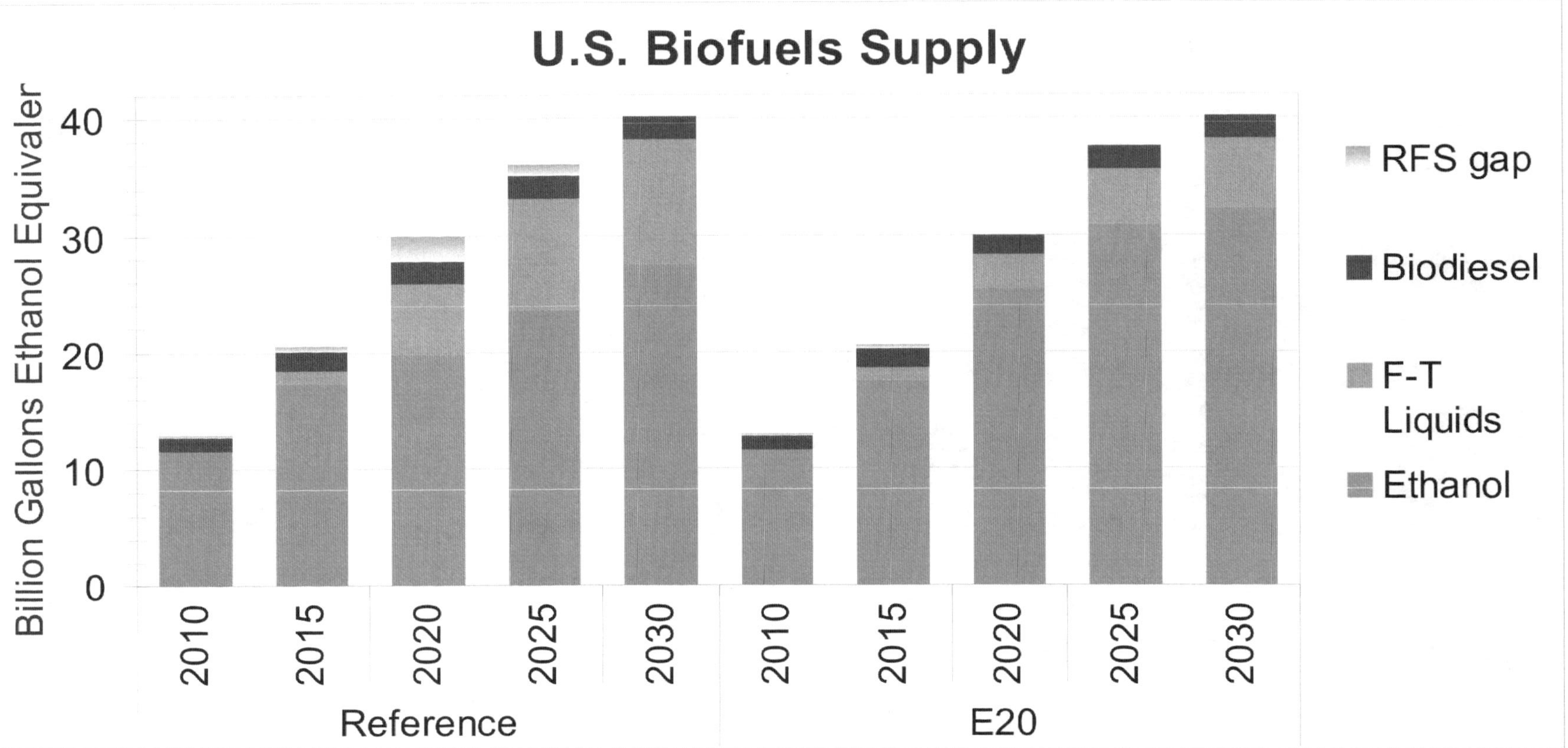

- **Only case to meet RFS**
- **Illustrates E85 infrastructure constraints**
  - **Pipelines, fueling stations, flexible fuel vehicles**

E20 Scenario: U.S. Supply Shares

- Significant increase in ethanol use.
- E20 allows lower cost ethanol to replace some F-T liquids and compliance credits (gasoline).
- E20 case shows benefits to reduce ETOH distribution constraints (e.g., expanded E85 retail outlets & more fuel-flexible vehicles).

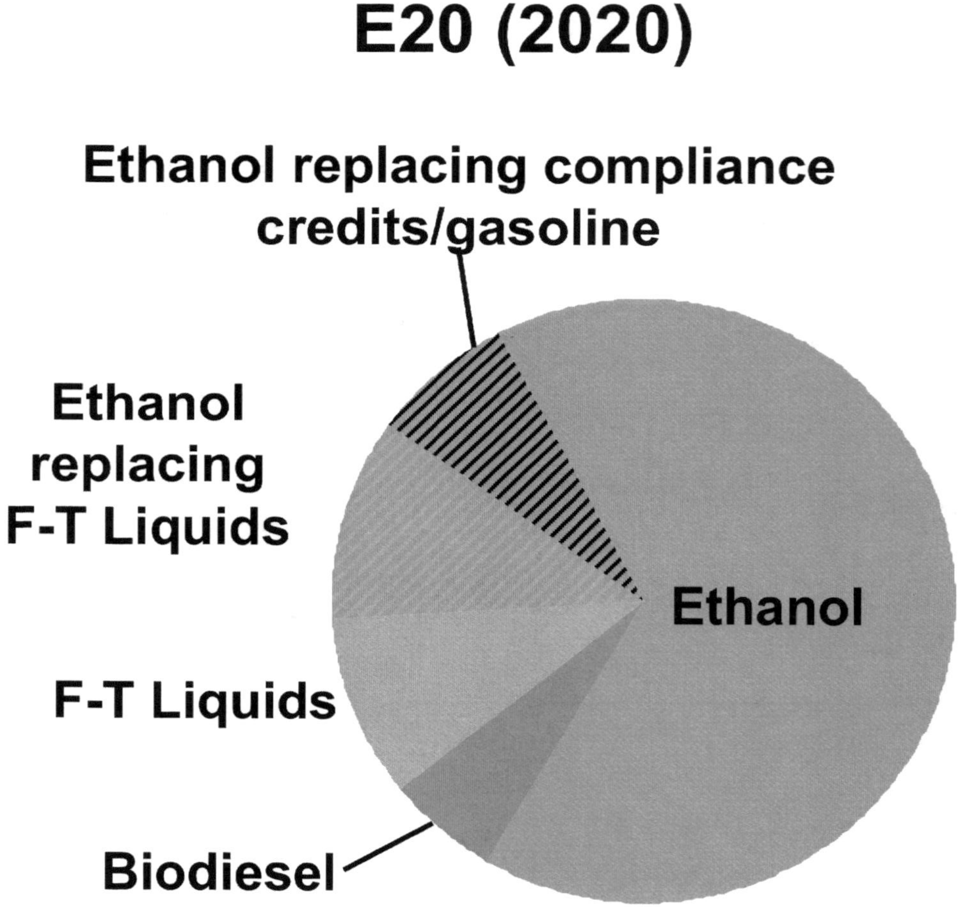

Total: 28 B gallons in Ref
30 B gallons in E20

E20 Scenario: U.S. Supply Shares

- Increase in ethanol is partly made possible by imports
- Imports increase by 60%

Reference (2020)

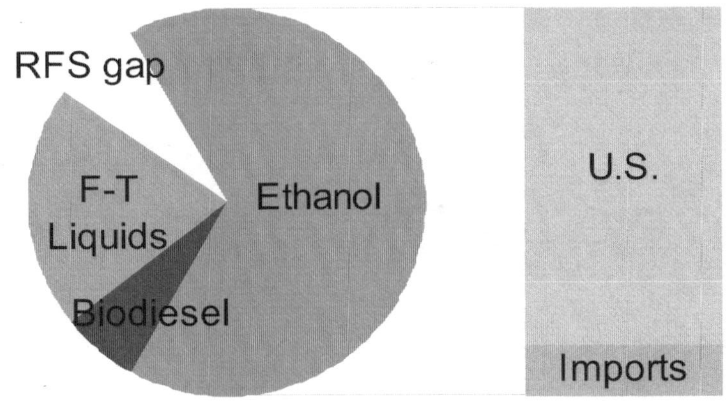

E20 (2020)

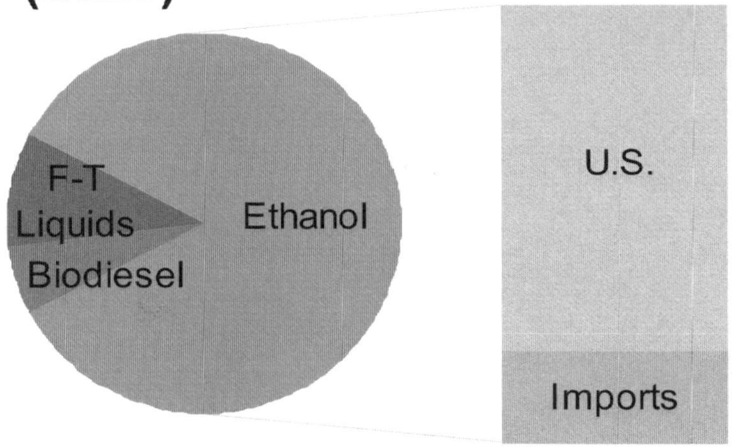

Conclusions

- Cellulosic biofuels are crucial share of RFS
  - Importance of learning investment and technology penetration
- E85 infrastructure constraints
  - Demonstrated by E20 scenario
  - Switch between biochemical and Fischer-Tropsch cellulosic
- Large volumes mandated, production is at inelastic portion of feedstock supply curve
  - Additional subsidies have little impact
- Sizeable role of imports (sugar and cellulosic)
- Implicit global price on CO_2, decline in grain ethanol
- High oil price, lower exports to U.S.

World Biofuels Study (WBS)

Collaboration

Project Management by **Office of Policy and International Affairs**

With Funding Support from **EERE / Office of Biomass Programs**

Feedstock Resource Potential

Conversion Process

Integrated Assessment

ORNL & NREL reports at http://www.osti.gov/bridge/ search 924080, 921804

267b Methane Gas Hydrate Equilibrium Inside Cylindrical Pores and Its Implications for Gas Production

Patricia Taboada-Serrano and Costas Tsouris. Oak Ridge National Laboratory, P.O. Box 2008, MS-6181, Oak Ridge, TN 37831

Clathrate hydrates are solid crystalline structures formed by water hydrogen-bonded cages. Each cage is stabilized via dispersion forces exerted by a gas molecule trapped in its interior. Gas hydrates are stable at moderate pressures and low temperatures, above the freezing point of water. Natural gas hydrates can be found in coastal sediments or the permafrost, and their exploitation constitutes an attractive future energy source. It is agreed within the scientific community that natural gas hydrate deposits contain more energy than all the other fossil fuel sources combined. Usually, natural gas hydrate deposits are present in the form of small disseminated particles in marine sediments. Currently, gas hydrate equilibrium models in the bulk are used to estimate the size of hydrate deposits in the ocean floor and investigate gas-production strategies. However, differences between estimates of the depth of the hydrate stability zone within marine sediments and acoustic measurements point towards the importance of the effects of confinement on hydrate thermodynamic stability. A thermodynamic model for the equilibrium of methane gas hydrate inside cylindrical pores was developed in an attempt to better predict the hydrate stability zone within marine sediments (i.e., depths at which natural gas hydrate is stable). The effects of the porous medium on hydrate equilibrium are included in the model in terms of surface energies. Predictions on the displacement of the hydrate stability zone in terms of pore size and other sediment characteristics will be presented and discussed. Accurate prediction of natural gas hydrate equilibrium in sediments is important for the design of gas harvesting strategies.

267d Novel Algal Photobioreactor Design and Operation for the Biofuels Industry

Nadia J. Abunasser[1], Mario Enrique Danton Garcia Perez[2], Haiying Tang[3], Steven O. Salley[3], John Wilson[4], and K. Y. Simon Ng[3]. (1) Chemical Engineering and Materials Science, Wayne State University, 461 Burroughs, Detroit, MI 48202, (2) Alternative Energy Technology, Wayne State University, 461 Burroughs, Detroit, MI 48202, (3) Department of Chemical Engineering and Materials Science, Wayne State University, 5050 Anthony Wayne Drive, Detroit, MI 48202, (4) National Biofuel Energy Laboratory, NextEnergy, 461 Burroughs ST, Detroit, MI 48202

The rising cost of traditional feedstock for the biofuels industry has led to an increasing interest in algae as a possible supplemental if not replacement feedstock. One of the main hurdles of this becoming a reality is the costs associated with growing, harvesting and extracting oil and carbohydrates from the algae. Here the cost of growing and harvesting the algae is addressed with a novel external loop airlift reactor. A reactor was designed to maximize the light penetration into the interior of the reactor in order to operate at the optimal optical density for the algal growth. This design includes the use of submersible fiber optics and a procedure for partial harvesting and addition of growth medium to maintain the required optical density. The flow characteristics in the reactor were also studied and modified to reduce the shear stress effects on algal death. The operation of the reactor is automatically controlled to maintain not only the appropriate optical density, but also the dissolved CO2 and dissolved O2 concentrations, pH and temperature for the algal strain being grown.

INCORPORATION OF REACTION KINETICS INTO A MULTIPHASE, HYDRODYNAMIC MODEL OF A FISCHER TROPSCH SLURRY BUBBLE COLUMN REACTOR

Donna Post Guillen, Anastasia M. Gribik, and Jonathan K. Shelley
Donna.Guillen@inl.gov, Anastasia.Gribik@inl.gov, Jonathan.Shelley@inl.gov
Idaho National Laboratory
PO Box 1625
Idaho Falls, ID 83415-3710, USA

Steven P. Antal
antals@rpi.edu
Department of Mechanical, Aerospace, and Nuclear Engineering
Rensselaer Polytechnic Institute
110 8th Street
Troy, NY 12180-3590, USA

ABSTRACT

This paper describes the development of a computational multiphase fluid dynamics (CMFD) model of the Fischer Tropsch (FT) process in a Slurry Bubble Column Reactor (SBCR). The CMFD model is fundamentally based which allows it to be applied to different industrial processes and reactor geometries. The NPHASE CMFD solver [1] is used as the robust computational platform. Results from the CMFD model include gas distribution, species concentration profiles, and local temperatures within the SBCR. This type of model can provide valuable information for process design, operations and troubleshooting of FT plants. An ensemble-averaged, turbulent, multi-fluid solution algorithm for the multiphase, reacting flow with heat transfer was employed. Mechanistic models applicable to churn turbulent flow have been developed to provide a fundamentally based closure set for the equations. In this four-field model formulation, two of the fields are used to track the gas phase (i.e., small spherical and large slug/cap bubbles), and the other two fields are used for the liquid and catalyst particles. Reaction kinetics for a cobalt catalyst is based upon values reported in the published literature. An initial, reaction kinetics model has been developed and exercised to demonstrate viability of the overall solution scheme. The model will continue to be developed with improved physics added in stages.

INTRODUCTION

As part of the Idaho National Laboratory's (INL's) Secure Energy Initiative, the INL is performing research in areas that are vital to ensuring clean, secure energy for the nation. Specifically, our team is developing a computational multiphase fluid dynamics (CMFD) model of Fischer Tropsch (FT) synthesis in a slurry bubble column reactor (SBCR) used for the production of synthetic fuels from secure, domestic resources. The model is being developed using a staged approach, starting with a simple model then becoming progressively more complex. Reaction kinetics have been incorporated into a hydrodynamic model for churn-turbulent flow using the NPHASE CMFD code as the robust numerical solver [1, 2]. The purpose of this model is to help industry understand the complex physical processes occurring in the SBCR. The model can be used as a numerical testbed to study custom problems that concern industry and gain a better understanding of the mechanisms that govern reactor performance from a more fundamental perspective than has been done in the past. To date, much of the understanding of these types of reactors has been obtained by limited studies using correlations that are not widely applicable over a range of operating conditions and scales. A more fundamental, mechanistic-based model that is broadly applicable is needed to enable efficient reactor design studies and scaling over the range from laboratory to pilot reactors and finally commercial production scales. To make the process economically viable, the designer must optimize reactor performance by varying operating conditions eliminating dead zones within the reactor. An understanding of the mixing behavior is essential for proper design and scale-up [3]. Simulations can be used to locate regions of maximum and minimum velocity and temperature, to facilitate identification of these dead zones and/or regions of excessively high temperatures, as well as predict any mass or heat transfer design limitations for the reactor. More sophisticated physicochemical models offer the potential to bring an increased level of understanding for the basic reaction and transport processes occurring within the reactor, with the eventual goal of being able to simulate the entire FT process and the associated plant [4]. Such knowledge will enable advances in process technology that are necessary for companies to remain world-class and competitive by enabling higher productivity, reducing pollution, using less energy and resources, and improving product quality.

REACTION KINETICS AND HYDRODYNAMICS MODEL DEVELOPMENT

An NPHASE CMFD model of the FT process in a SBCR is under development. The SBCR is a multiphase chemical reactor where a synthesis gas, comprised mainly of H_2 and CO, is bubbled through a liquid hydrocarbon wax containing solid catalyst particles to produce specialty chemicals, lubricants, or liquid fuels. The FT synthesis reaction is the polymerization of methylene groups [-(CH_2)-] forming mainly linear alkanes and alkenes, ranging from methane to high molecular weight waxes. The FT process is described by the set of exothermic reactions described by Eqs 1-4

$$CO + 3H_2 \rightarrow CH_4 + H_2O \qquad \Delta H_{298} = -247 \text{ kJ/mol} \qquad \text{(Eq 1)}$$

$$nCO + 2nH_2 \rightarrow -(CH_2)-_n + nH_2O \qquad \text{(Eq 2)}$$

$$H_2O + CO \rightarrow CO_2 + H_2 \qquad \Delta H_{298} = -41 \text{ kJ/mol} \qquad \text{(Eq 3)}$$

$$2CO \rightarrow C + CO \qquad \Delta H_{298} = -172 \text{ kJ/mol} \qquad \text{(Eq 4)}$$

The selection of catalyst (i.e., avoiding Ni) can minimize the production of methane, the reaction shown in Eq 1. The chain polymerization kinetics in Eq 2 produces a hydrocarbon product distribution described by the Anderson-Schulz-Flory (ASF) model. A broad spectrum of mainly alkanes and alkenes with carbon numbers from C_1 to C_{50+} are produced. Process conditions and/or catalyst selection can be used to tailor the product distribution [5]. Eq 3, the water-gas-shift (WGS) reaction, is invoked only for reactions involving Fe-based catalysts. Eq 4 is the Boudouard reaction resulting in the deposition of carbon and the formation of carbon dioxide.

The computational platform for this work is the NPHASE computer program, an unstructured, finite volume, multifield, pressure based computational multiphase fluid dynamics (CFD) computer code offering both segregated and coupled numerical solution methods. To achieve numerical convergence for multiphase flows, simulations are performed using the robust, coupled algorithm, which solves for the phasic velocity, pressure, and volume fraction simultaneously. Although, the computer memory requirements are increased, this ability to couple the conservation of mass and momentum equations is more robust than the use of a segregated solver. A full three-dimensional, Eulerian-Eulerian framework is employed, rather than tracking a very large number of individual bubble trajectories. Ensemble-averaged conservation equations for mass, momentum, energy, interfacial area, and species transport are solved for an arbitrary number of fields or phases and species. The equations are discretized into a block matrix system, which is solved by an algebraic multigrid solver. The fully coupled mass/momentum setup allows any field/phase to interact with any other field/phase within the multigrid matrix system. A detailed derivation of the ensemble-averaged conservation equations has been given by [6]. The ensemble-averaging approach models a time-varying complex multiphase flow using a well-defined theoretical framework. The option for invoking an additional transport equation that is solved for the interfacial area concentration [7] is available in the NPHASE code and may be included in the FT model to better predict the available reaction area between the gas and bulk fluid. Interfacial mass, momentum, turbulence and heat transfer sub-models provide coupling between the fields and phases. NPHASE also includes transport equations for turbulence modeling and a arbitrary number of species in any carrier field. Turbulence in the bulk fluid was modeled using a standard k-ε model with the two-phase turbulence viscosity given by [8]. The total kinematic viscosity for the bulk flow is the sum of molecular, shear-induced and bubble-induced components.

In an industrial FT process, syngas is sparged at the base of a large cylindrical vessel containing liquid and catalyst. Momentum is imparted to the bulk liquid via the kinetic energy of the sparged syngas and the recirculating internal flow is driven by interfacial forces. For the process to be economically viable, the velocity of the gas must be sufficiently high that the reactor operates in the churn-turbulent flow regime.

A key factor for properly capturing the underlying physics of two-phase flows is the formulation of consistent mechanistic closure laws that describe the dominant mass, momentum and thermal interactions at the fluid interfaces. Closure relations are needed to reintroduce the information that was lost as a result of averaging the conservation equations. These closure laws account for the sub- scale mass and momentum transfer between the various fields and phases. The interfacial interactions between the individual fields are specified by mechanistic models for both drag and non-drag forces. The interfacial momentum transfers considered important for churn-turbulent flow in a bubble column include the interfacial drag force, the virtual mass force, the wall force, the lift force, and the turbulence dispersion force. Refinement of these mechanistic models is ongoing to ensure that they apply over a realistic range of operating conditions and scales. The closure relations include coefficients that need to be tuned to the physics of the flow regime. Optimization software (Engineous iSIGHT-FD) is being used to perform a sensitivity study by varying the ranges of the coefficients over their theoretically appropriate ranges.

The gas field is divided into a small (5 mm) bubble group and a large (50 mm) slug/cap bubble group. Each bubble group is treated as a separate field to allow different flow physics to be included via the closure models. The model also accounts for interactions between bubble groups (i.e., breakup and coalescense). Incorporation of a third bubble size (< 1 mm) will also be evaluated f in the model, since these very small bubbles present a large interfacial area available for the reaction process, relative to the bubble volume. Eventually multiple bubble groups, at least 5 groups, will be considered for future simulations.

Bubble dynamics play a key role in the transport phenomena and in the overall rates of reaction. The intrafield mass transfer between the small bubbles and slug/cap bubbles was formulated using coalescence and breakup models described by [9], [10] and [11]. Improved bubble fragmentation and coalescence sub-models are an area where more research is needed. Two opposing theories have been proposed regarding the roles of the large versus the small bubbles on mass transfer. One theory espouses the concept that large cap bubbles largely determine the flow pattern, whereas the small bubbles (assumed to be spherical) govern mass transfer. Another theory claims that the unstable, oscillating surface of the large bubbles create a high level of turbulence, which enhances mass transfer, whereas the small bubbles have a relatively rigid surface that inhibits mass transfer. Studies by [12] found that bubble fragmentation and coalescence were strongly influenced by the turbulence dissipation rate of the liquid phase.

This section presents the model formulation, governing equations and closure laws used to simulate the turbulent, four-field, two-phase flow for a bubble column operating in the churn-turbulent flow regime (see Figure 1). Figure 2 outlines the thermal fluid and chemistry features to be incorporated into the current NPHASE CMFD model. The SBCR model will continue to evolve and improve as the best available physical sub-models are incorporated. The local instantaneous conservation equations are averaged and phasic volume fractions are introduced.

The phasic continuity equation is written as (field j for phase k)

$$\frac{\partial}{\partial t}\left(\alpha_{jk}\rho_{jk}\right)+\nabla\cdot\left(\alpha_{jk}\rho_{jk}\underline{u}_{jk}\right)=m'''_{jk} \qquad \text{(Eq 5)}$$

The phasic momentum equation is written as (field j for phase k)

$$\frac{\partial}{\partial t}\left(\alpha_{jk}\rho_{jk}\underline{u}_{jk}\right)+\nabla\cdot\left(\alpha_{jk}\rho_{jk}\underline{u}_{jk}\underline{u}_{jk}\right)=-\alpha_{jk}\nabla p_{jk}+\nabla\cdot\alpha_{jk}\left(\underline{\underline{\tau}}_{jk}+\underline{\underline{\tau}}_{jk}^{Re}\right)+\alpha_{jk}\rho_{jk}g+\underline{M}_{jk}+m'''_{jk}\underline{u}_{i} \qquad \text{(Eq 6)}$$

The terms on the right hand side of the equation represent the pressure gradient, shear stress tensor, body force due to gravity, momentum exchange at interfaces and momentum flux due to mass transfer.

The compatibility condition included in the coupled solver equation set, is expressed as

$$\sum_{k=1}^{n}\alpha_{k}=1 \qquad \text{(Eq 7)}$$

The phasic energy equation is written as

$$\frac{\partial}{\partial t}\left(\rho_{jk}h_{jk}\right)+\nabla\cdot\left(\rho_{jk}\underline{u}_{jk}h\right)=\nabla\cdot\left(q_{jk}''+q_{jk}''^{T}\right)+\Delta H_{i}+h_{c}A(T_{liq}-T_{w}) \qquad \text{(Eq 8)}$$

The terms on the right hand side of the equation represent the conductive heat flux, the volumetric heat source due to the chemical reaction, and convective heat transfer to the heat exchanger, respectively. The energy equation is used to predict the change in liquid temperature due to the exothermic chemical reaction and heat exchanger. A simple heat exchanger model has been used which allows heat transfer from the liquid and represented as a constant temperature heat exchanger over the central region of the column. Future work should incorporate a more detailed model of the heat exchanger tubes within the SBCR including the pressure drop due to the presence of internals.

The physical phenomena in the SBCR occur at different time scales. Due to the thermal inertia of the reactor and its contents, the heat transfer scale is many orders of magnitude larger than the mass transfer/reaction time scale [13]. The chemical reaction rate is used to evaluate the rate of change in the mass fraction of the species. The local mass fractions of the chemical species are tracked using seven separate species transport equations. The reaction rate is used to evaluate the local heat addition to the bulk flow. The species transport equation can be written as

$$\frac{\partial}{\partial t}\left(\alpha_{jk}\rho_{jk}Y_{s}\right)+\nabla\cdot\left(\alpha_{jk}\rho_{jk}\underline{v}_{jk}Y_{s}\right)=\nabla\cdot\alpha_{jk}\left(\frac{\mu_{jk}}{\sigma_{s}}+\frac{\mu_{jk}^{T}}{\sigma_{s}^{T}}\right)\nabla Y_{s}+S_{s} \qquad \text{(Eq 9)}$$

where Y_s is the species mass fraction and S_s is the source term containing the reaction kinetics

$$S_s = W_s \sum_{n=1}^{N_R} n_{sn} R_n \qquad \text{(Eq 10)}$$

Chemical reactions are being added in stages of increasing complexity. The first stage of the model tracks the mass fraction species of H_2 and CO in the gas phase and [-(CH_2)-] in the liquid field and includes absorption of gas species from both large and small bubbles into the bulk liquid phase. The driving force for the gas across the interface into the bulk liquid will be dependent upon the interfacial species concentration in both small and large bubbles. However, because it is difficult to measure the concentration at the gas-liquid interface, coefficients for convective mass transfer across an interface for the overall driving force between the bulk concentrations in the gas and liquid phases are implemented. The product from the incorporation of absorption will be the steady state concentration profile of the absorbed gas species in the bulk liquid phase and the impact of the hydrodynamics on the concentration profile. All phases are assumed to have a constant density. The variation of gas density with temperature and pressure will be accounted for in future simulations.

In this initial model we assumed that the small catalyst particles are sufficiently small such that external and internal mass and heat transfer are not rate limiting. A macrokinetic approach using power rate laws is used, where the heterogeneous catalytic reaction is described by the following rate equations

$$-R_{CO} = \frac{a c_{CO,L} c_{H_2,L}}{\left(1 + b c_{CO,L}\right)^2} \qquad \text{(Eq 11)}$$

$$-R_{H_2} = 2\left(-R_{CO}\right) \qquad \text{(Eq 12)}$$

$$R_{H_2O} = -\left(-R_{CO}\right) \qquad \text{(Eq 13)}$$

$$R_{-(CH_2)-} = -\left(-R_{CO}\right) \qquad \text{(Eq 14)}$$

with the kinetic constant *a* and the adsorption coefficient *b* defined as follows [14]

$$a = 8.852 \cdot 10^{-13} \exp\left[4494.41 \frac{1}{K}\left(\frac{1}{493.15} - \frac{1}{T}\right)\right] (R\,T)^2 \frac{m^6}{kg_{cat} s\, mol} \qquad \text{(Eq 15)}$$

$$b = 2.226 \cdot 10^{-5} \exp\left[-8236.15 \frac{1}{K}\left(\frac{1}{493.15} - \frac{1}{T}\right)\right] R\,T \frac{m^3}{mol} \qquad \text{(Eq 16)}$$

The reaction model utilizes the macrokinetic rate expression for a cobalt catalyst developed by Yates and Satterfield. A cobalt catalyst was selected for the initial model, since the WGS reaction can be neglected. The intrinsic kinetic expression for the consumption of CO and H_2, which is a Langmuir-Hinshelwood type expression, was based on data collected over a range of industrially relevant conditions. The local mass fractions of the chemical species are evaluated using species transport equations. Seven species are currently tracked in the SBCR simulation: CO reactant in small bubbles, large bubbles and bulk liquid; H_2 reactant in small bubbles, large bubbles and bulk liquid; and the [-(CH_2)-] reaction product in the bulk liquid. Subsequent simulations will include a product representing the average properties of the liquid hydrocarbon wax mixture ($C_{27}H_{56}$) plus water. Eventually, the entire ASF product distribution will be modeled by grouping the products into light hydrocarbons, gasoline, diesel and waxes [5].

Change in moles of the reacting species, vapor-liquid equilibrium and the resulting temperature of the catalyst and fluid phases will be solved simultaneously. To simplify the model, it is initially assumed that only CO and H_2 are absorbed into the liquid phase and vapor liquid equilibrium can be neglected. Later, gas species products of the reaction will be added to the adsorption model, including H_2O, CH_4, CO_2, C_2H_4, C_2H_6, and other light gas species, as well as vapor liquid equilibrium. Solubility will be computed from correlations developed from experimental gas solubility data for n-paraffin solvents [15].

MODEL SETUP

A constant inlet gas superficial velocity of 60 cm/s was specified uniformly over the entire bottom of the reactor vessel. The bulk liquid mass flow rate is 1.15 kg/s. The inlet gas flow was assumed to be comprised of 25% small bubbles (5 mm) and 75% large bubbles (50 mm). The inlet mass fraction was 0.7 CO and 0.3 H_2. A 10% solids loading with 75 μm catalyst particles was suspended in the liquid.

A length-to-diameter ratio (L/D) of 4.5 was used for the simulation. For subsequent work, the L/D will be increased to more closely resemble an industrial reactor. A 60 x 500 grid with refined cell size near the wall, the inlet and the free surface was used to resolve the recirculating flow field expected in the SBCR.

The free surface in the gas disengagement region near the top of the column is represented as a constant pressure surface through which gas can escape, but the liquid and catalyst can not penetrate. The liquid and catalyst can flow out of the domain via one layer of nodes along the side to allow overflow to escape the reactor.

For this analysis, the convergence criteria are based upon an average root-mean-square (RMS) change in velocity, pressure and volume fraction between iterations being less than a specified tolerance. The RMS change in velocity was less than 1.0×10^{-4} while the pressure RMS change was less than 1.0×10^{-2}. These levels are low enough that further reducing the error would not change the predicted hydrodynamic profiles.

RESULTS AND DISCUSSION

Results from the CMFD model are presented below. Color contour plots illustrating the spatial distribution of various computed quantities are illustrated in Figures 3 through 10. The volume fraction of small bubbles, large bubbles and catalyst are shown in Figure 3. The radial distribution of gas holdup is higher at the center of the column and exhibits a power-law profile as seen in experimental data by [16]. The distribution of catalyst within the SBCR impacts the efficiency of the chemical processes and thus the productivity of column as a chemical reactor. The slug/cap bubble interfacial lift force causes a lateral migration of the larger bubbles to the central region of the column. In this simulation, no lift force was applied to the small bubbles but future simulation will include an interfacial lift force. Experimental results show that the smaller bubbles move towards the wall [17] even in churn-turbulent flows. Future simulations will apply a lift force to both bubble groups. Contours of CO, H_2 and [-(CH_2)-] concentration are shown in Figure 4. The concentration of reactants is highest in the lower portion of the SBCR, whereas the product concentration is highest at the upper portion of the SBCR. This is a result of the reactants transformed into product as the syngas travels through the reactor. Figure 5 depicts the velocity magnitude of bulk liquid, large bubbles and catalyst. Due to the small catalyst particle size, the catalyst particles track with the bulk fluid. Figure 6 provides the reaction rate for [-(CH_2)-] production, the temperature rise above inlet temperature due to the addition of heat by the exothermic reactions and the static pressure. The pressure is seen to vary as the hydrostatic pressure. The middle region of the SBCR is at the lowest temperature in the column, since this is the region over which the heat exchanger was applied. Figure 7 shows the small and large bubble CO concentration, and small and large bubble H_2 concentration. Figure 8 shows the diffusion rate of CO from small and large bubbles into the bulk liquid. Figure 9 gives the small and large bubble H_2 diffusion rate into the bulk liquid. Figure 10 shows the computed bulk fluid turbulent kinetic energy and catalyst loading. The results of parametric studies performed using the NPHASE model indicate that it is necessary to further investigate and validate turbulent dispersion in the model.

The reader is cautioned that these results are shown to demonstrate the capabilities of the NPHASE code and have not been fully validated with data. Limited validation of the hydrodynamic model has been accomplished, but the results incorporating heat transfer and reaction kinetics have not. The hydrodynamics of an air/water bubble column flow computed with the NPHASE code have been compared to experimental data for liquid velocity and gas holdup in a previous study [1]. Agreement with experimental data was good and the code predicted the major trends exhibited by the flow. It is stressed that the results must be validated with high-quality data suitable for validating a fundamentally-based CMFD code. Much existing experimental data has been obtained for use as input to lower order models (such as the one-dimensional axial dispersion model) or to develop empirical correlations. Thus, data for global parameters (overall gas fraction or pressure drop) is much more common than that for local parameters (spatial distribution of phases, phasic velocity, etc.).

Computational simulations presented here can be used in concert with experiments to enhance our understanding of the physicochemical processes occurring within a FT SBCR. Suitable data sets for validating predictions of chemically reacting flow in a bubble column are being sought. The model can be used as an effective means for designing experiments.

SUMMARY

A four-field, seven-species ensemble-averaged multifluid CFD model is under development to model the flow, thermal effects and chemical concentrations within a FT SBCR. The CMFD model is fundamentally based which allows it to be applied to different industrial processes and reactor geometries. The NPHASE CMFD solver is used as the robust computational platform. Results obtained include phase distribution, velocity and species concentration profiles throughout the reactor. Additional demonstration cases and a preliminary validation of the model are underway. Suitable data to validate the model under relevant conditions (pressure, temperature, etc.) are being pursued. Future work is also needed to validate the hydrodynamics predictions and extend the present simplified model to more realistic modeling assumptions and conditions.

We are working towards the goal of multiphase, science-based computational simulations that play a significant role in the design, operation and troubleshooting of FT plants. It is our goal to develop a more fundamental, mechanistic-based model that is more broadly applicable than currently available, and can scale to prototypic sizes and operating conditions. Recent advances in CMFD modeling tools coupled with the availability of massively parallel computing platforms offers the potential realization of this goal.

ACKNOWLEDGMENT

Funding for this work was provided by Battelle Energy Alliance, LLC under Contract No. DE-AC07-05ID14517 with the U.S. Department of Energy.

NOMENCLATURE

ASF	Anderson-Shulz-Flory
CFD	computational fluid dynamics
CMFD	computational multiphase fluid dynamics
FT	Fischer Tropsch
INL	Idaho National Laboratory
RMS	root-mean-square
SBCR	slurry bubble column reactors
WGS	water gas shift

a	unitless	kinetic constant
b	unitless	adsorption coefficient
$c_{i,G}$	mol/m^3	bulk concentration of species i in the gas phase
$c_{i,L}$	mol/	bulk concentration of species i in the liquid phase
c_i^*	mol/m^3	equilibrium concentration of species i in the gas phase with the liquid
$D_{L,i}$	m^2/s	diffusivity for species i in the liquid phase
$D_{L,ref}$	m^2/s	reference diffusion coefficient in the liquid
h	kJ/kg	enthalpy
h_c	$W/m^2 \cdot K$	heat transfer coefficient
ΔH	kJ/mol	heat of reaction
$k_L a_i$	$s^- m^{31}$	overall mass transfer coefficient for species i
m'''	$kg/s \cdot m^3$	interphase mass transfer rate per unit volume
M_{jk}	$N/s \cdot m^3$	interphase momentum transfer rate per unit volume
n	unitless	carbon number
n_{sn}	unitless	species s in reaction n
p	Pa	pressure
q	W/m^2	heat flux
R	J/mol-K	gas constant
$R_{-(CH2)-}$	$mol/kg_{cat}s$	rate of production for $-(CH_2)-$
R_{CO}	$mol/kg_{cat}s$	rate of consumption for CO
R_{H2}	$mol/kg_{cat}s$	rate of consumption for H_2
R_{H2O},	$mol/kg_{cat}s$	rate of production for H_2O
R_i	mol/m^3	rate of mass transfer for the solute (gas) into the solvent (liquid)
S_s	unitless	source term
T	K	temperature
$\underline{V}$	m/s	velocity
Y_s	unitless	species mass fraction
W_s	kg/mol	molecular weight of species s
x_n	unitless	mass fraction

α	unitless	void fraction
ρ	kg/m^3	density
μ	kg/m·s	viscosity
τ	N/m	shear stress

Subscripts

cat	catalyst
i	interfacial
j	field
k	phase
liq	liquid
w	wall

REFERENCES

1. Antal, S.P., R.T. Lahey, Jr., M.H. Al-Dahhan, *Simulating Churn-Turbulent Flows in a Bubble Column using a Three-Field, Two-Fluid Model*, in *5th International Conference on Multiphase Flow, ICMF'04*. 2004: Yokohama, Japan.
2. Gribik, A., Guillen, D.P., Ginosar, D., *Kinetic Modeling of a Fischer-Tropsch Reaction over a Cobalt Catalyst in a Slurry Bubble Column Reactor for Incorporation into a Computational Multiphase Fluid Dynamics Model*, in *International Pittsburgh Coal Conference 2008*. 2008, September 29 – October 2, 2008: Pittsburgh, PA.
3. Steynberg, A.P., and Dry, M.E., *Fischer-Tropsch Technology*. Studies in Surface Science and Catalysis, ed. G. Centi. Vol. 152. 2006: Elsevier.
4. Guillen, D.P., Boardman, R., Gribik, A., Wood, R., Carrington, R. *Integrated Fischer Tropsch Modular Process Model*. in *Proceedings of the 49th Annual Meeting of the Idaho Academy of Science*. 2007. Idaho Falls, ID.
5. Bartholomew, C.H., Farrauto, R.J., *Fundamentals of Industrial Catalytic Processes*. Second Edition ed. 2006, Hoboken, NJ: John Wiley & Sons, Inc. p. 402-404.
6. Drew, D.A., Passman, S.L.. "Theory of Multicomponent Fluids", *Applied Mathematical Sciences*, **Vol. 135**, (1998)
7. Murakawa, H., Antal., S.P., and Lahey, Jr., R.T. in *Proceedings of the 2008 International Congress on Advances in Nuclear Power Plants (ICAPP)*. 2008. Anaheim, CA.
8. Sato, Y., Sekoguchi, K., "Liquid Velocity Distribution in Two-Phase Bubble Flow", *Int. J. Multiphase Flow*, **Vol 2**: p. 79-95, (1975)
9. Antal, S.P., M.Z. Podowski, R.T. Lahey, Jr., D. Barber, C. Delfino, *Multidimensional Modeling of Developing Two-Phase Flows in a Large Adiabatic Riser Channel*, in *The 11th International Topical Meeting on Nuclear Reactor Thermal-Hydraulic (NURETH-11)*. 2005: Popes Palace Conference Center, Avignon, France.
10. Podowski, M.Z., Antal, S.P., Tiwari, P., Wierzbicki, B.Z., "Development of Mechanistic Models of Two-Phase Flows for the NPHASE Code," Rensselaer Polytechnic Institute, Report to the U.S. Nuclear Regulatory Commission, (2004).
11. Luo H., Svendsen, H.F., "Theoretical Model for Drop and Bubble Breakup in Turbulent Dispersions", *AIChE Journal*, **Vol. 42**, (1996)

12. Krepper, E., Lucas, D., Frank, T., Prasser, H.-M., Zwart, P.J., "The Inhomogeneous MUSIG Model for the Simulation of Polydispersed Flows", *Nuclear Engineering and Design*, **238**: p. p. 1690-1702, (2008)
13. Mahecha-Botero, A., Grace, J.R., Elnashaie, S.S.E.H., Lim, D.J., "Time-Scale Analysis of a Fluidized-Bed Catalytic Reactor Based on a Generalized Dynamic Model", *The 12th International Conference on Fluidization - New Horizons in Fluidization Engineering*, **RP4**(76): p. p. 623-628, (2008)
14. Yates, I.C. and Satterfield, C.N., "Intrinsic Kinetics of the Fischer-Tropsch Synthesis on a Cobalt Catalyst", *Energy & Fuels*, **5**: p. 168-173, (1991)
15. Chao, K.C., Lin, H.M., "Synthesis Gas Solubility in Fischer Tropsch Slurry: Final Report," Purdue University, School of Chemical Engineering, DOE/PC/70024-T9 (1988).
16. Rados, Novica, *Slurry Bubble Column Hydrodynamics*, in *Department of Chemical Engineering*. 2003, Washington University: Saint Louis, Missouri.
17. Lucas, D., Krepper, E., Prasser, H.-M., Manera, A., "Investigations on the Stability of the Flow Characteristics in a Bubble Column", *Chem. Eng. Technol.*, **29**(9): p. p. 1066-1072, (2006)

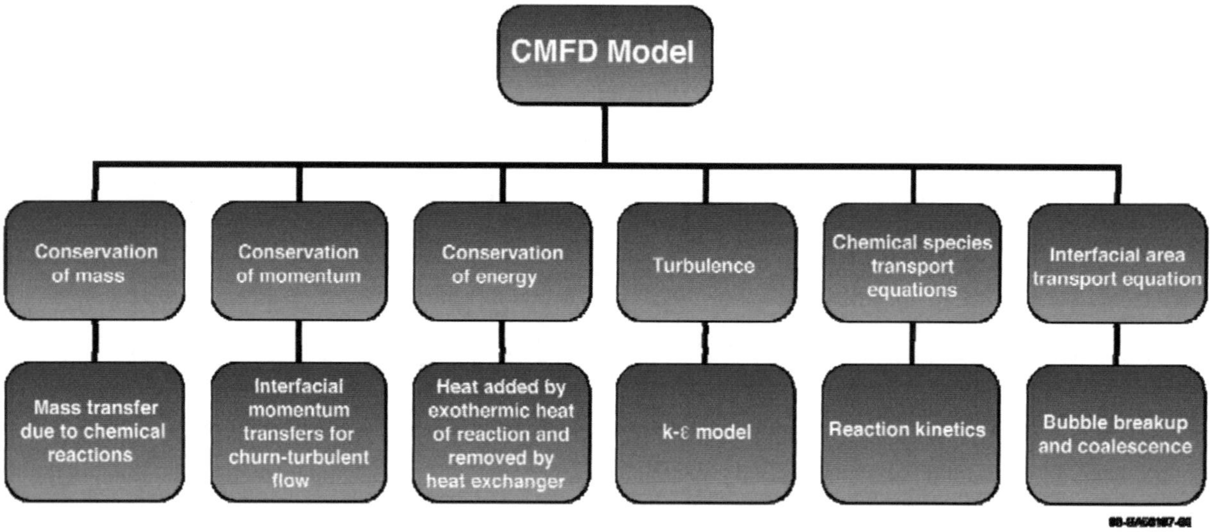

Figure 1. Overview of CMFD FT model.

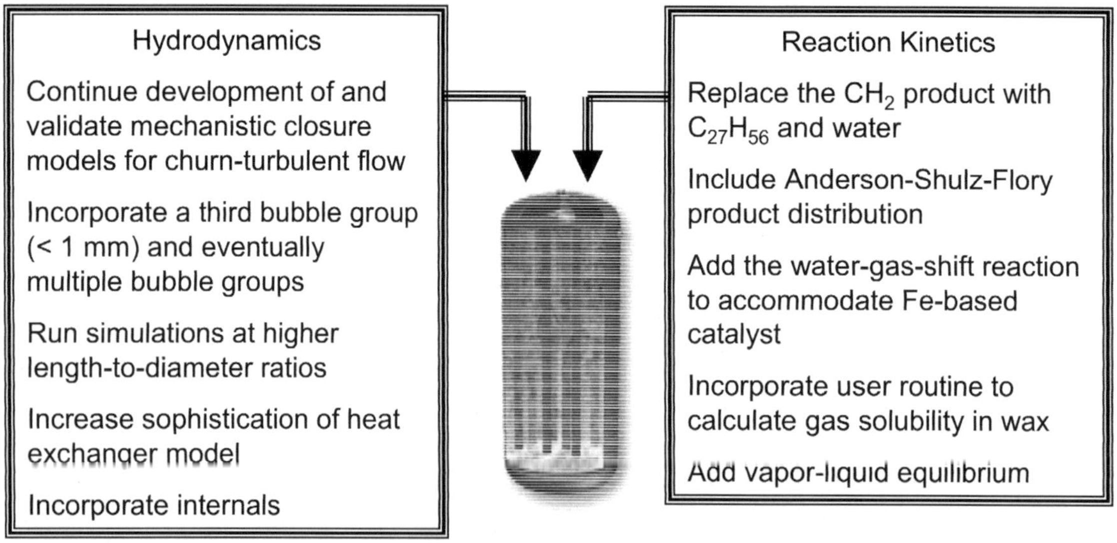

Figure 2. Staged approach to incorporating physicochemical features into preliminary model.

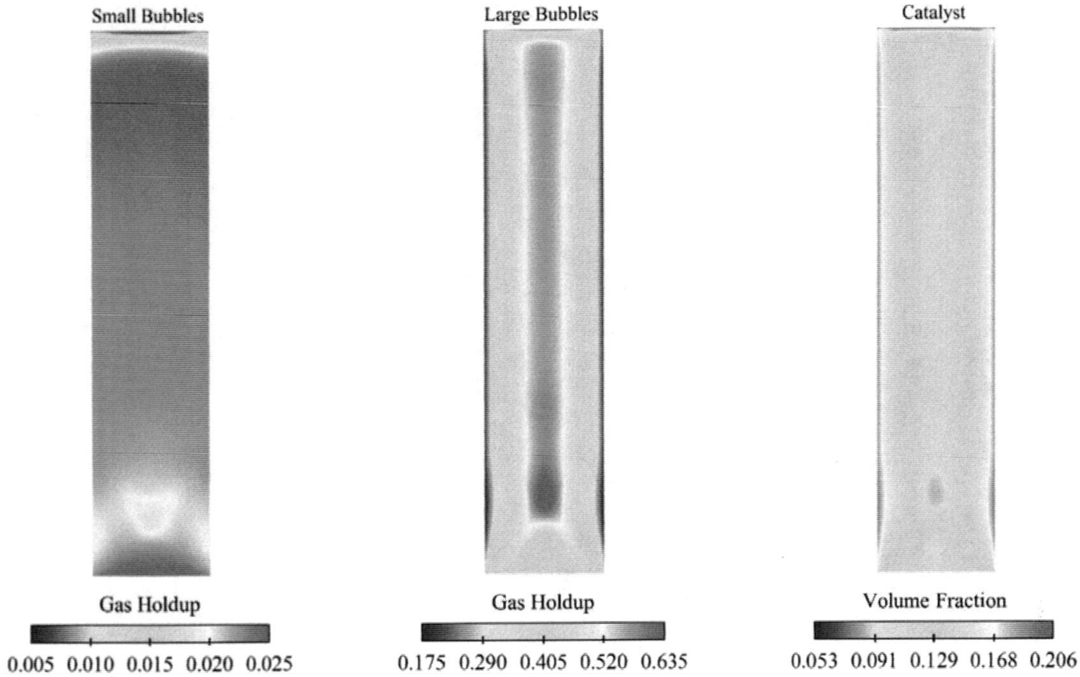

Figure 3. Volume fraction of small bubbles, large bubbles and catalyst.

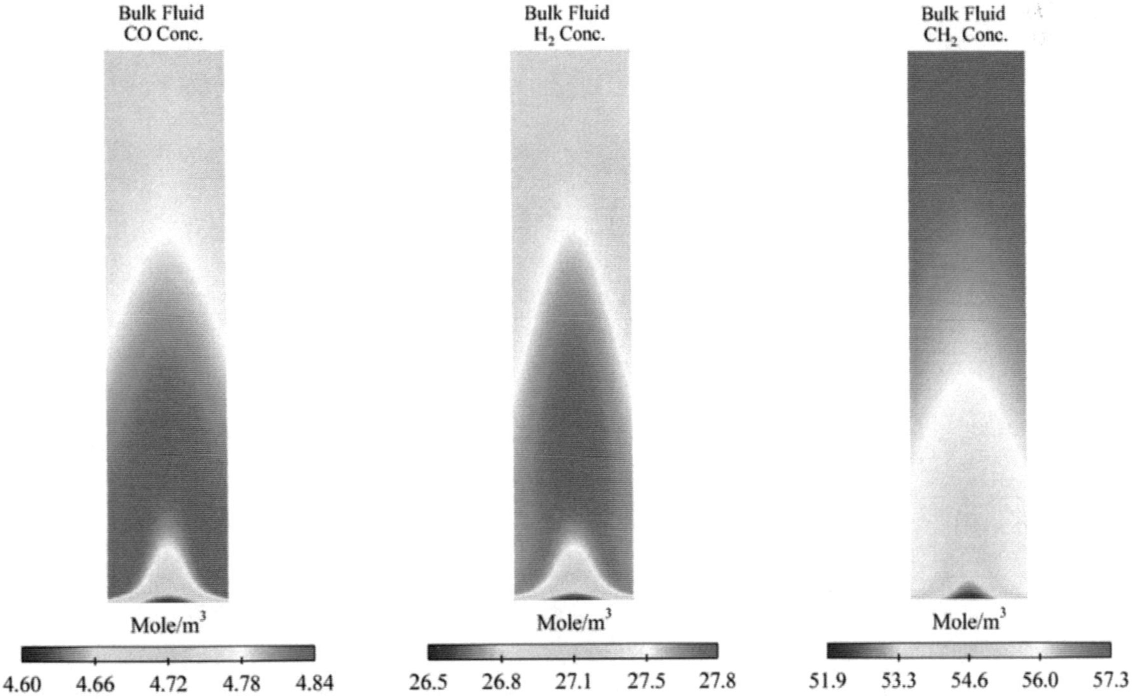

Figure 4. CO, H_2 and [-(CH_2)-] concentration.

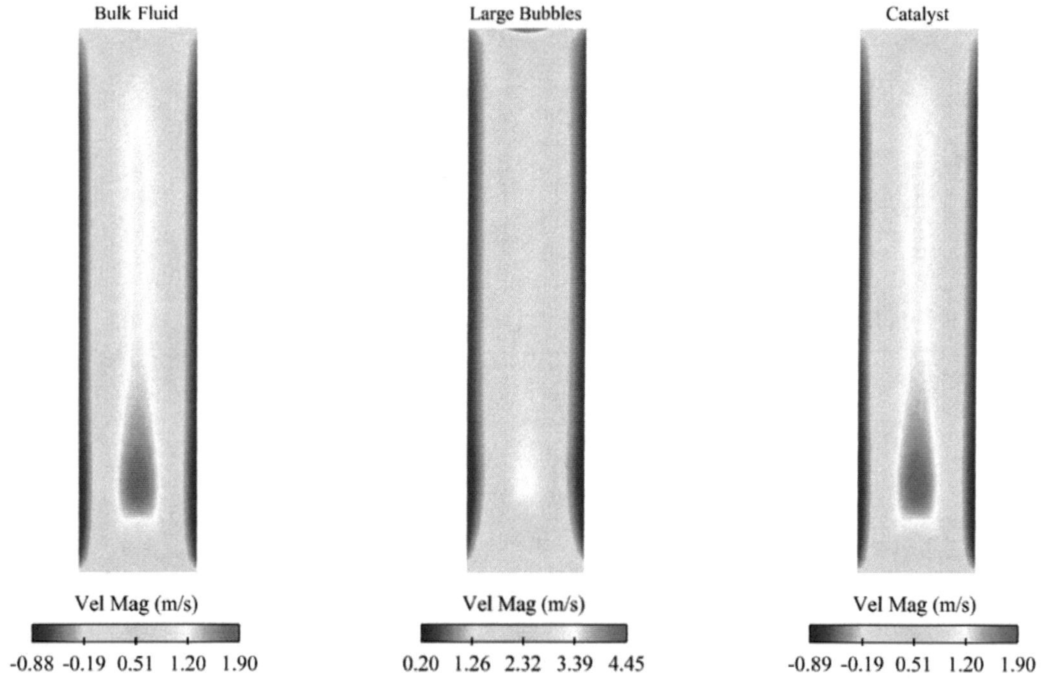

Figure 5. Velocity magnitude of bulk liquid, large bubbles and catalyst.

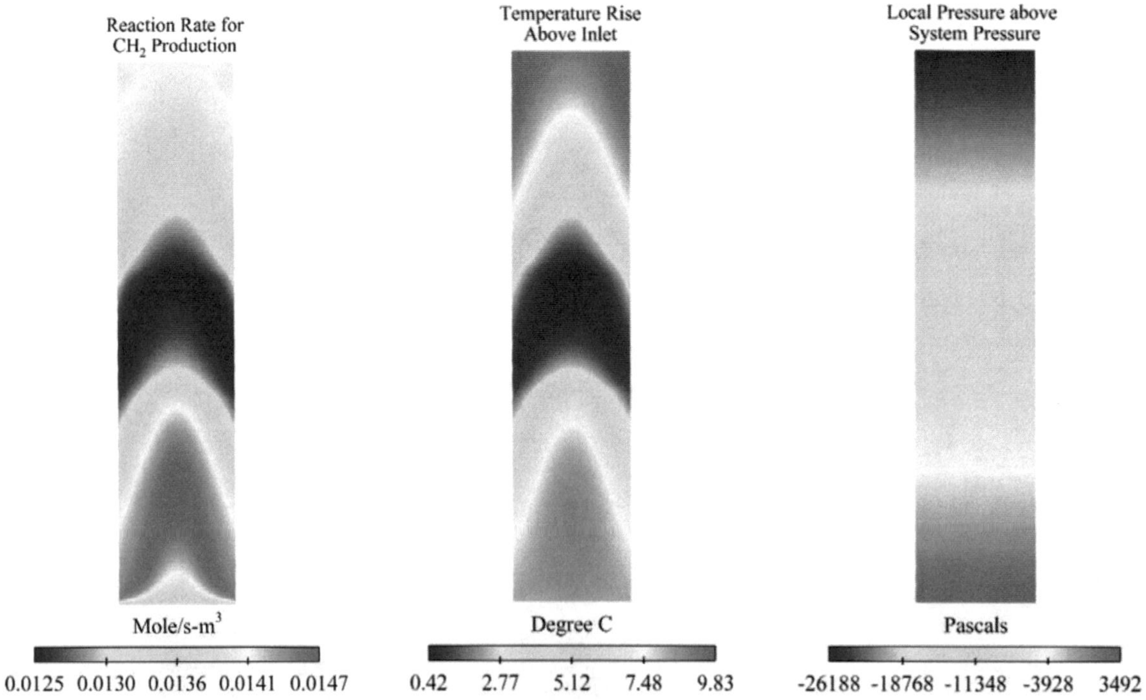

Figure 6. Reaction rate for [-(CH_2)-] production, temperature rise above inlet temperature and static pressure.

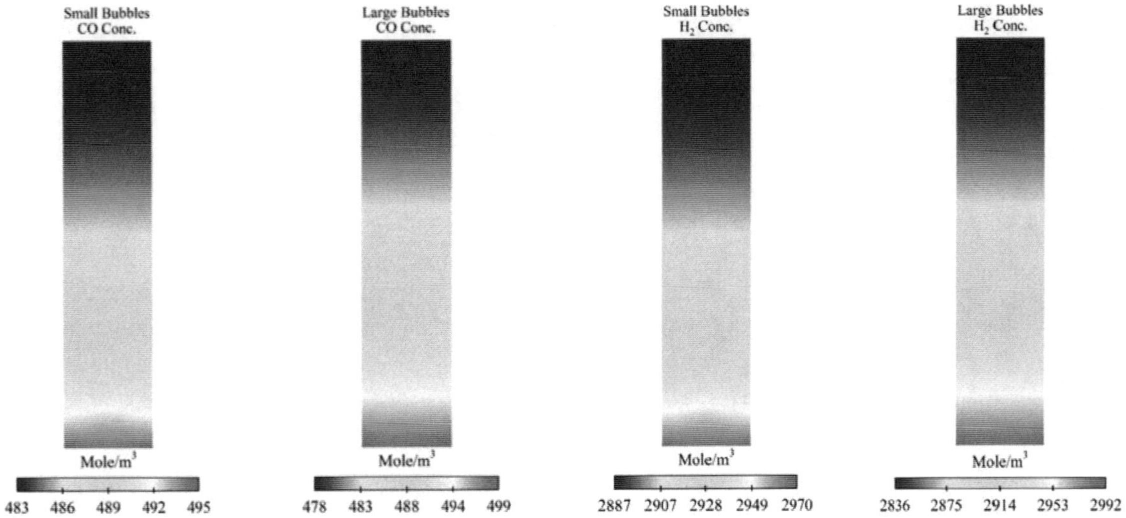

Figure 7. Small bubble CO concentration, large bubble CO concentration, small bubble H_2 concentration and large bubble H_2 concentration.

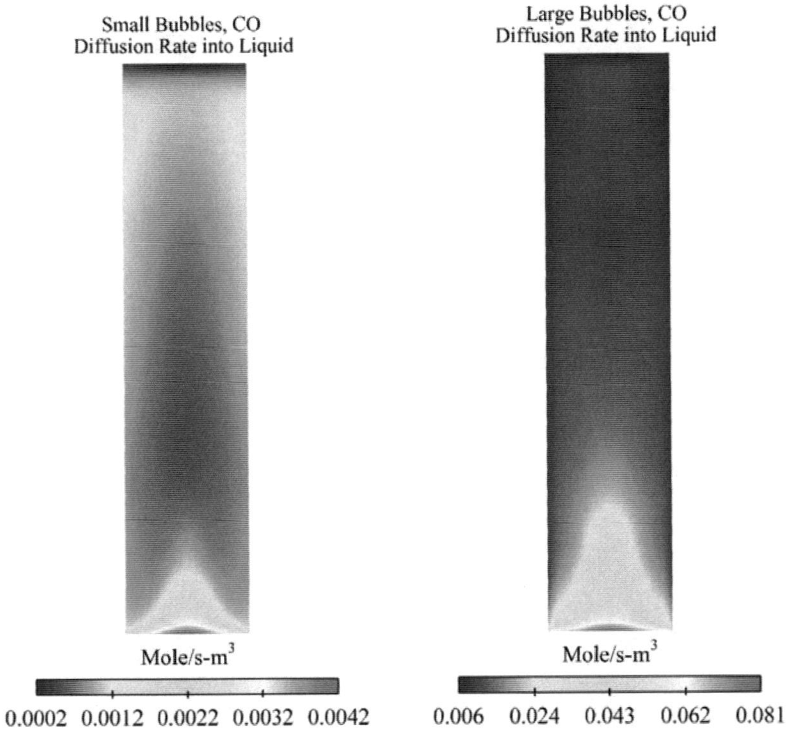

Figure 8. Diffusion rate of CO from small and large bubbles into liquid.

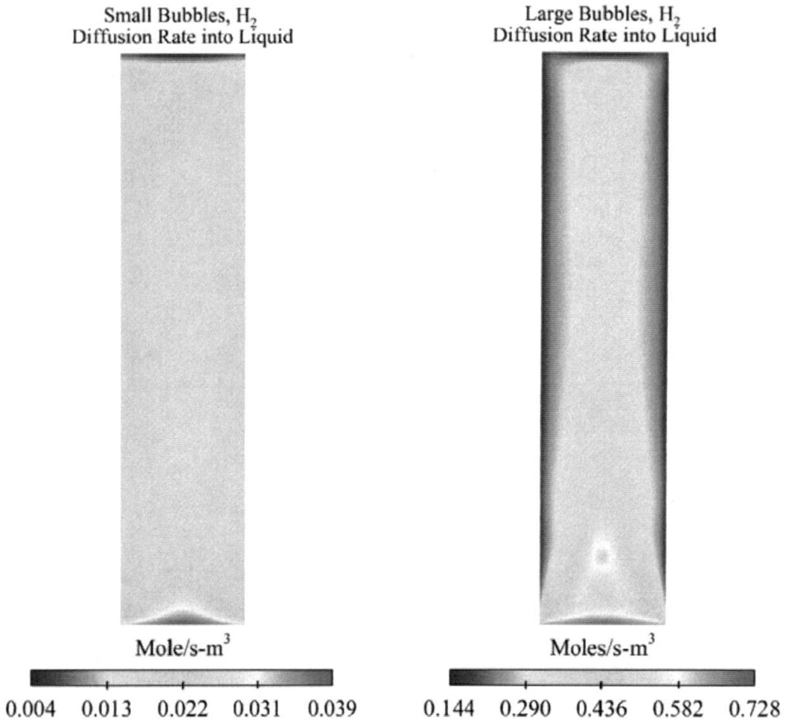

Figure 9. Small and large bubble H_2 diffusion rate into liquid.

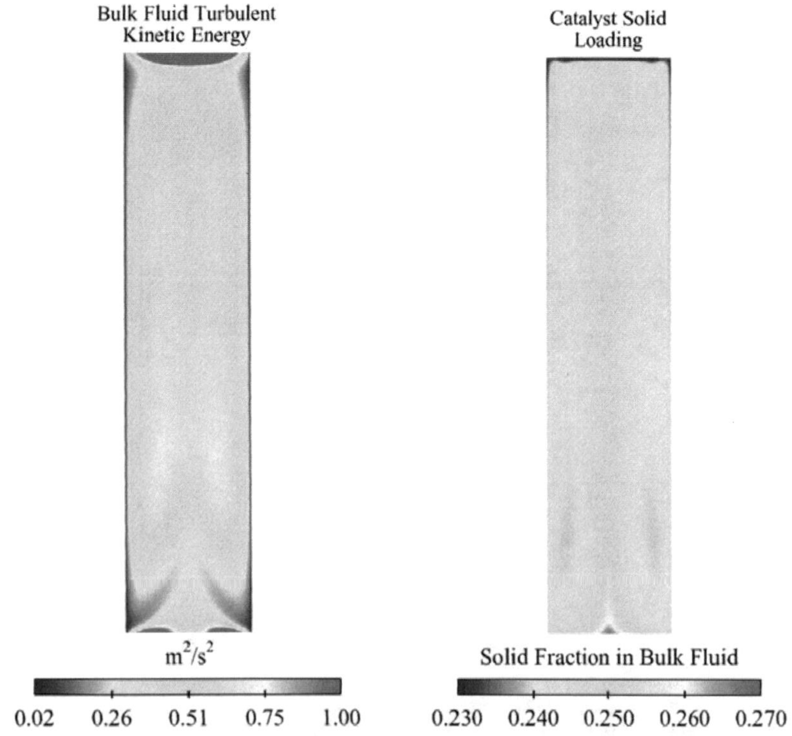

Figure 10. Bulk fluid turbulent kinetic energy and catalyst loading.

267f Oleaginous Microorganisms Utilizing Glycerol as Sole Carbon Source for the Production of Lipids for Biofuels

J. Mathew Thomas, Rafael Hernandez, Todd French, William Holmes, Earl Alley, and Jaclyn Hall. Dave C. Swalm School of Chemical Engineering, Mississippi State University, 323 Presidents Circle, Swalm Chemical Engineering, Mississippi State, MS 39762

The growing popularity of biodiesel as an alternative to petroleum-derived diesel fuel has created a dramatic increase in the production of crude glycerol. Currently crude glycerol has a low market value and many biodiesel producers are treating the crude glycerol as a waste product. If biodiesel is to succeed as an alternative to petroleum-based fuels, it is essential that the crude glycerol produced be given value. Oleaginous microorganisms have the ability to use many different types of carbon sources and convert them into oils that can then be used for the production of fuels. The purpose of this study is to investigate the feasibility of oleaginous microorganisms to utilize glycerol as a carbon source for the production of oils. The study will focus on several different species of oleaginous yeasts and bacteria. Kinetics of glycerol removal and cell growth will be presented, as well as quality and yield of oil from the oleaginous microorganisms.

330a Global View of Technical Aspects of Fuel Ethanol Production

Charles D. Tereck, KATZEN International, Inc., 2300 Wall Street, Suite K, Cincinnati, OH 45212

As the World increases its commitment to renewable energy demand, motor fuel grade ethanol (MFGE) production has been growing at a rapid pace and represents the largest bio-based, renewable fuel. The selection of feedstock is discussed as it relates to availability and processing technology employed in different regions around the world. Various technologies and processing schemes are presented that facilitate the efficient conversion of the selected feedstock to ethanol and by-products.

330b Accelerating Sample Preparation in the Biofuel R&D by Automated High Output Technologies

Josef G. Schroer, Chemspeed Technologies, 7 Deer Park Drive, Suite L, Monmouth Junction, NJ 08852

The general need for increased efficiency in the laboratory has kick-started the development of innovative equipment for workflow automation.

Many laboratories within the chemical industry in the meantime have adopted the use of automated High-Output-Technologies, which has been boosting productivity in the pharmaceutical industry since several decades.

New research fields like "Biofuels & Renewable Energy Sources" recently started to automate their workflows to keep up with an ever-increasing amount of samples.

For this reason instruments similar to the equipment, which originally has been designed to accelerate the process of Drug Discovery, now commonly are being used for parallel Sample Preparation, Pretreatment, Digestion and Analysis.

Multiple dispensing of solid and liquid samples, sample treatment with corrosive chemicals under harsh conditions, work-up steps like filtration & SPE purification are only some of the most relevant workflow steps.

Other technical requirements to these instruments might strongly differ from workflow to workflow. Generally a diverse set of complex unit operations needs to be automated, which are characteristic to each individual workflow.

For this reason only flexible, highly modular, and scalable equipment has the chance of covering the complex automation needs of this emerging field.

Using selected examples, from the "Biofuels & Renewable Energy" R&D, this presentation will show how a variety of challenging parallel workflows fully have been automated.

WindFuels – Competitive Transportation Fuels From Wind Energy and Waste CO_2

Dr. F. David Doty, Dr. Laura L Holte, Glenn N Doty, David McCree, Siddarth Shevgoor
Doty Energy, Columbia, SC, USA. August, 2008 www.WindFuels.com

Introduction

It is essential for the market to help drive the dramatic cut needed in CO_2 emissions to prevent a climate disaster in this century. We show that the economics for producing clean liquid hydrocarbons and alcohols from water and waste CO_2 on wind farms improved by an order of magnitude between 2002 and 2008, and we present the scientific and technological basis for another factor-of-two improvement in these economics. Wind energy is by far the most competitive renewable energy resource in many regions. The perceived challenge is getting wind energy from good sites to where and when it is needed, both for the transportation sector and for the power grid. Efficient conversion of wind energy and waste CO_2 into clean, stable, liquid fuels – also called *WindFuels* – solves these problems. Annual *WindFuels* production per land area in good wind regions should exceed biofuels production density in fertile farming areas by a factor of five.

First, water and renewable electrical power are fed into an electrolyzer, which produces the required hydrogen. Then waste CO_2 (from coal power plants) and the renewable hydrogen are piped into a novel Reverse Water Gas Shift (RWGS) process, which permits practical reduction of CO_2 to the CO needed at efficiencies approaching theoretical limits. The renewable CO and H_2 are then fed into a novel Renewable Fischer-Tropsch Synthesis (**RFTS**) process. There they are converted into mostly mid-alcohols (ethanol, propanol, and butanol) and other valuable chemicals. These fuels may then be readily stored and distributed by conventional means – pipelines and tanker trucks. The electrolyzer also produces a huge amount of oxygen, which may be sold if market conditions warrant, or it may be utilized in novel processes to improve the plant efficiency.

From basic thermodynamics, one readily calculates that the theoretical maximum chemical efficiency of synthesis of ethanol from H_2 and CO_2 is 80.1% without utilization of the excess heat released from the Fischer Tropsch Synthesis (**FTS**) reactor. If this excess heat is also converted in an ideal heat engine, the theoretical combined-cycle efficiency limit is ~90%. One advantage of ethanol may be appreciated by noting that its synthesis from H_2+CO_2 results in 1.5 molecules of water per carbon atom in the fuel, while the synthesis of alkanes or alkenes results in 2 molecules of waste water per carbon atom in the fuel.

Previously, fossil-based FTS efficiencies for environmentally attractive fuels, such as ethanol and propanol, have usually been under 40%. The need for very efficient recycling of the large amounts of H_2, CO, and CO_2 in the byproducts from mid-alcohols FTS has possibly been the strongest argument against mid-alcohols FTS compared to gasoline, lubricants, and diesel. A novel plant design is presented, validated by many simulations, that permits order-of-magnitude reduction in energy penalties associated with the major separations in a fully recycled mid-alcohols plant compared to previously published designs. The example wind-driven RFTS plant size chosen for illustration and simulated in detail assumes 250 MW ***mean*** input electrical power. It achieves **72%** FTS-plant higher heating value (**HHV**) efficiency in production of mid-alcohols and other products from H_2 and recovered (waste) CO_2, or about **60%** net HHV efficiency when including the electrolyzer at expected near-term performance. At least eight separate, substantial innovations in the system design combine to permit this major advance in RFTS efficiency and cost effectiveness.

It is worth pointing out that as recycle efficiency and byproduct utilization both approach 100%, which is the primary thrust in the RFTS optimization, catalyst selectivity and activity both become much less important. Catalysts never affect the thermodynamics – only the kinetics of the various pathways. Hence, further development of catalysts will have limited effect on system efficiency in a plant with 95% recycle efficiency, though of course it will influence profitability by changing the product mix.

The annual U.S. demand for the various chemicals that are not major fuel components that would come from the RFTS reactors (free of sulfur, salts, metals, halides, and nitrogen) is nearly 100 million tons, and this highly profitable market exceeds 140 billion dollars. The most economically attractive route to a major reduction in green-house gases (**GHGs**) appears to be synthesis of liquid fuels and chemicals from wind energy and waste CO_2. These global markets can support the rapid scale-up needed to avert a climate disaster in this century.

Overview of the RFTS Plant Design

The simplified flow diagram depicted in **Figure 1** is useful for presenting a plant overview and system summary. The ***first key*** to competitive performance is obtaining pressurized high-purity hydrogen and oxygen at high efficiency, which in turn requires operating an electrolyzer at very high pressure. Preheated water [121] is fed into the alkaline electrolyzer [123] that is powered by renewable electricity [122] to produce the oxygen and hydrogen. The pressurized O_2 and H_2 are then optimally expanded before being used. The source hydrogen, at ~4 MPa (near term), further heated using waste heat, is then expanded in a turbo-generator [125] to ~1 MPa. The cleaned, source CO_2 is heated and expanded in another turbo-generator [126]. Both gases are then further heated [127] before being fed into the RWGS reactor [128].

The ***second key*** is a major advance in cost-effectiveness of gas-to-gas heat recuperation, which is essential for efficient implementation of both the RWGS and the FTS reactions. The importance of approaching isothermal conditions in heat transfer has been understood for many decades, but methods of doing so in gas-to-gas recuperators have had limited success. Our fundamentally new approach to recuperator design is expected to permit an order of magnitude improvement in cost-effectiveness of recuperators of high thermal effectiveness. This recuperator is the subject of a pending patent application, and more details will be presented at a later time.

The ***third key*** is efficient reduction of CO_2 to CO. Two viable approaches – denoted as "multi-stage RWGS" and "recycle RWGS" – are presented. To drive the reaction equilibrium to the right, most of the water must be efficiently condensed out of the RWGS products [129] as the reaction progresses, so clearly ultra-high-performance gas-to-gas recuperation [130] is central to either approach. The diagram of Figure 3 is somewhat more representative of Multi-stage RWGS than of Recycle RWGS, though the latter should ultimately be preferred. In the recycle case, a $CuAlCl_4$-aromatic complexing method is used to separate the CO and drive the reaction even farther to the right. If there is excessive CO_2 in the RWGS products, it needs to be recycled [132]. The CO and H_2 from the RWGS reactor are then compressed in a turbo-compressor [133] to produce the pressurized "new syngas" [134], with typical molar-% compositions as noted in the diagram. This is combined [135] with the preheated recycled syngas [147] and fed into the FTS reactor [140]. A fixed-bed multi-tubular FTS reactor design is shown to have advantages for high-pressure, variable-rate, low-conversion, high-temperature, highly exothermic reactions, as needed for high yield of mid-alcohols.

The ***fourth key*** is achieving dramatically improved efficiency in handling low-conversion FTS processes by using high-pressure condensers [141] for the initial separations. Further compression [142] to 8-14 MPa may be needed to achieve adequate gas and product separations in cryogenic condensers [143]. To achieve adequate FTS-catalyst lifetime, it is necessary to separate much of the WGS-CO_2 [144] from the FTS products for re-conversion to CO in the RWGS reactor. A novel boost-expand separation process is presented that allows nearly an order of magnitude lower power consumption than more common CO_2 separation methods. This is possible partly because of efficient cryogenic recuperation [147] of the cooling capacity in the recycled syngas after its expansion in a turbine [146] back to the pressure needed in the FTS reactor. The separation also benefits from the recuperator advances mentioned previously, and it benefits markedly from higher FTS reactor operating pressure – a counter-intuitive discovery.

The ***fifth key*** is designing a plant that is inherently compatible with operation over a very wide range of mass flow rates. Variable-angle nozzles, variable-speed motors and generators, and turbine switching assist to this end, along with storage of the gases at several temperatures and pressures, and the use of optimal heat transfer processes. Numerous additional features further improve efficiency, including a refrigeration cycle utilizing the free compressed oxygen, a dual-source organic Rankine cycle heat engine, and an improved CH_4 separation process.

The ***sixth key*** is simplified local upgrading because of the absence of troublesome impurities in the crude products and because of the availability of abundant hydrogen, oxygen, low-grade waste heat, electrical power, and excess cryocooling capacity. Other beneficial aspects of the separations processes allow simplified recovery of all flash gases and eliminate any significant purge stream. The ***seventh key*** is efficiently converting the waste heat from the two major sources (the electrolyzer and the FT reactor) to electricity. This advance is the subject of another pending patent application.

It is important to appreciate that the most important factor influencing the advanced RFTS plant design optimization is the rapidly changing costs in different forms of energy. In 2002, for example, the average cost of electrical energy to the U.S. industrial user was about \$15/GJ, while the cost of bulk gasoline was about \$6/GJ. The average cost of grid-quality electrical energy on wind farms in favorable regions in 2012 will probably be about \$13/GJ (similar to its current cost), while the mean cost of *bulk* gasoline will likely be over \$35/GJ (\$4.00/gal) – even without a carbon tax.

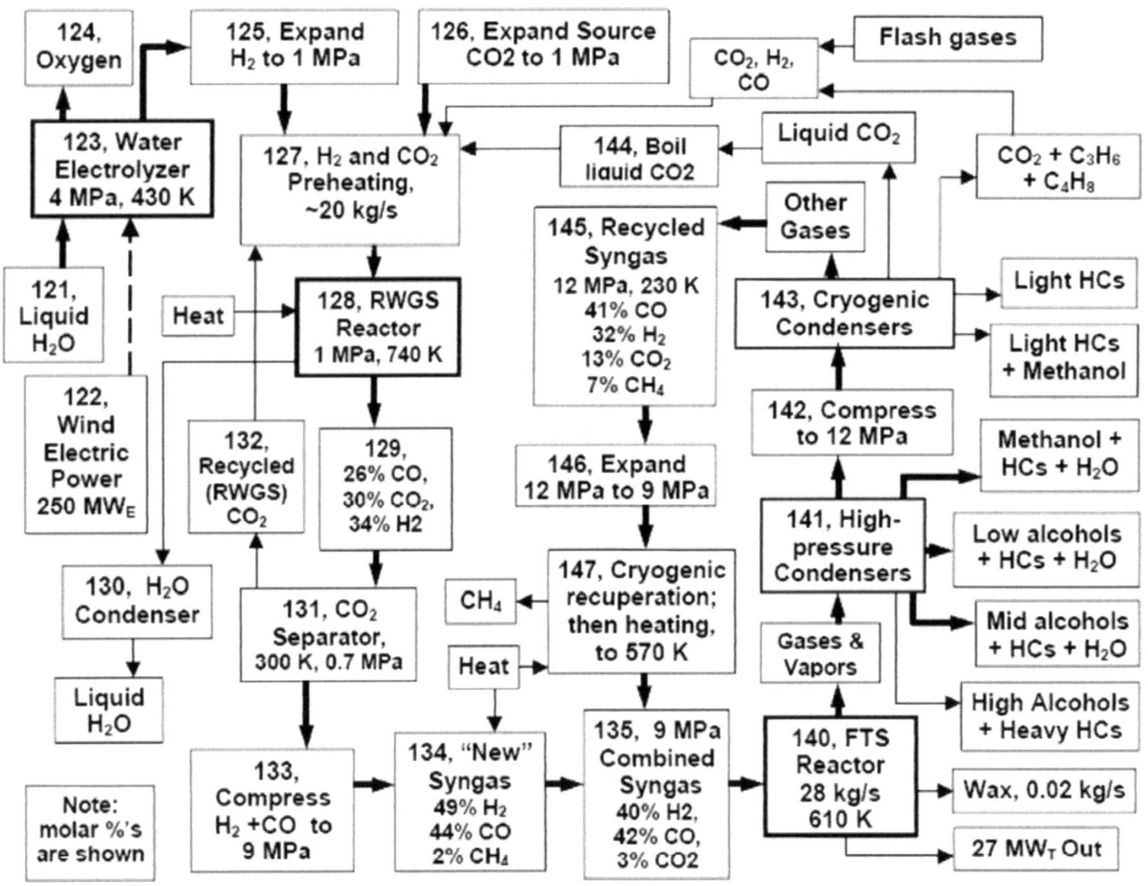

Figure 1. Simplified RWGS RFTS Plant Schematic.

Efficient, Renewable, Syngas Production. Efficient syngas production begins with efficient hydrogen production. Electrolyzers for efficiently splitting water into high purity hydrogen and oxygen have been in industrial production for decades. A concentrated solution of potassium hydroxide (KOH) in water is used to minimize resistivity and support ion mobility. The addition of four electrons to four water molecules at the negative electrode (cathode) produces two molecules of hydrogen gas ($2H_2$) and four hydroxyl ions ($4OH^-$), which remain in the solution. At the positive electrode (anode), four electrons are removed from four OH^- ions, producing two molecules of water ($2H_2O$) and a molecule of oxygen (O_2). A membrane that is permeable to the OH^- ions separates the two electrodes to keep the gases from mixing while allowing the electrical current to flow through it on the charge carriers. (In practice, the solutions on both sides are continually flowing to maintain the desired salt concentrations.) The two gases produced also contain a lot of water vapor (and its enthalpy, which must be utilized effectively) but only minute traces (easily under 1%, and sometimes under 0.01%) of other impurities.

Efficiency of commercially available 2 MW electrolyzers has typically been 73% higher heating value (HHV), but this efficiency measure has not included the mechanical energy of the compressed gases, which adds 2% at 4 MPa. Laboratory experiments have exceeded 85% HHV at higher pressures and lower current densities, and recent progress in catalysts for both the anodes and cathodes show promise for further gains in HHV efficiency. We have shown that the waste heat (at 160 ^{0}C) can be utilized at over 30% efficiency. Hence, total system efficiency of a 250 MW electrolysis system may eventually approach 90%.

The next step is to efficiently get the carbon monoxide needed in the syngas from CO_2. A number of electrolysis processes have been demonstrated, but their efficiencies at practical current densities have been low, and they have utilized very expensive components and have not demonstrated the needed reliability.

A very robust and efficient reaction, known for more than a century as the water gas shift reaction (WGS), is commonly used for the production of hydrogen (and unwanted CO_2) from carbon monoxide and water vapor, usually at temperatures in the range of 400 to 800 ^{0}C. The reverse of this reaction, known as the reverse water gas shift (RWGS) provides a robust method of producing CO and water from CO_2 and H_2. Getting this reaction to achieve high yield of CO at high efficiency with low production of unwanted methane (CH_4) has previously been a challenge, but it now appears practical. Near term efficiency of this process should now exceed 85%, and 92% may eventually be achieved.

The endothermic RWGS reaction is given by the forward direction of:

$$CO_2 + H_2 \rightleftarrows CO + H_2O \ , \quad \Delta H = 38.9 \text{ kJ/mol.} \qquad [1]$$

The reverse of the RWGS, the WGS, is easy to achieve at low-temperatures (450-550 K) and high pressures using Cu/ZnO catalysts, but the needed low temperature RWGS has seen relatively little investigation and utilization. Generating syngas from $CO_2 + H_2$ has not been an objective of much prior work, due to the expense of H_2 from electrolyzed water compared to the cost of methane. Until now, the market has not had a well articulated need for an optimum low-temperature RWGS catalyst. The RWGS reaction has often been seen as an undesirable competing reaction to be suppressed – as in methanol synthesis.

There are several exothermic reactions competing with the RWGS:

$$CO_2 + 3H_2 \rightleftarrows CH_3OH + H_2O \ , \quad \Delta H = -61.5 \text{ kJ/mol,} \qquad [2]$$

$$CO_2 + 4H_2 \rightarrow CH_4 + 2H_2O \ , \quad \Delta H = -179 \text{ kJ/mol,} \qquad [3]$$

The following exothermic reactions and the exothermic reverses of eqs. [3] and [6] also compete when CO is present in sufficient amounts.

$$CO + H_2 \rightarrow C + H_2O \ , \quad \Delta H = -135 \text{ kJ/mol,} \qquad [4]$$

and

$$2CO \rightarrow C + CO_2 \ , \quad \Delta H = -174 \text{ kJ/mol.} \qquad [5]$$

At the pressures and temperatures appropriate to optimize the RWGS relative to methane, methanol production is usually negligible. Moreover, in synthesis of mid-alcohols, considerable CH_3OH may simply be fed into the FTS reactor, where it can be converted to mid-alcohols.

Carbon deposition – leading to catalyst deactivation – is usually dominated by the Boudouard reaction, eq. [5]. Its activation energy is rather high (113 kJ/mol), but it is critical that it not be catalyzed. Of course, reducing the CO partial pressure will quickly reduce the reaction rates of eqs. [4] and [5] and several other possible paths.

Some have thought that the RWGS reaction cannot be made to work adequately below 720 K, and this may be true at high reactor pressures (over 3 MPa) with high space velocities and low excess CO_2 and H_2. However, it is not difficult to accommodate excess H_2 and CO_2 in the product stream, low recuperator pressures, and moderate space velocity.

A catalyst with good selectivity to CO is helpful in keeping CH_4 down, especially at higher pressures. Some of the more effective RWGS catalysts for the 520-720 K range at 0.3 to 3 MPa (total pressure) include Au/TiO_2, Cu/silica, and Cu/alumina. Even higher selectivity (98%) and excellent activity have been reported for a 0.9%-Pt doped Ca/C catalyst at low temperatures, though it is rather expensive. With proper space velocity, it appears that methane and carbon production can be kept low by operating below 1 MPa H_2 partial pressure at lower temperatures (below 660 K), or below 0.3 MPa H_2 partial pressure at higher temperatures – at least to 700 K with Cu catalysts, and possibly to 970 K with Fe_3O_4/Cr_2O_3 or future catalysts. Most data thus far for practical conditions (well above atmospheric pressure, low methane yield, low-cost catalysts, low C deposition, and high CO yield) are also at rather low space velocity, so more development here could improve performance and reduce reactor cost.

The dominant limitation in the catalyzed RWGS reaction is the WGS reaction, as the reverse is always also catalyzed. The easiest way to reduce the WGS is to condense the water from the mixture after partial reaction – and then re-heat and repeat this cycle as necessary. This can permit high conversion of the CO_2 to CO (though CO_2 separation from the product will still be needed) with little additional energy penalty if highly effective counterflow heat exchangers are employed. Suitable, low-cost exchangers for moderate-pressure operation (which is critical for keeping methane low with metal catalysts) are not yet commercially available, but our recuperator breakthrough makes them practical.

The RWGS reaction may be driven further to the right by including CO removal from the products as the reaction progresses. Several methods for CO separation have been demonstrated. The most widely implemented is the COSORB method of Kinetics Technology International (originally developed by Tenneco Chemicals), which uses a solution of $CuAlCl_4$ in toluene for the selective absorption of CO from mixtures containing CO_2, H_2, CH_4, and inerts. The Cu(I)-CO complex is formed at about 300 K and moderately high pressures (0.3-3 MPa), and the CO is released at about 400 K and low pressures (0.1-0.5 bar).

Semi-permeable membranes and molecular sieves (such as zeolite 5A) are also available with fairly good selectivity for CO. All of these methods are more expensive than simple H_2O condensation and add significant gas compression penalties. However, the combination of CO and H_2O removal from the products may allow the RWGS reaction to work adequately below the FTS reaction temperature, and that benefit should eventually more than offset the costs associated with CO separations.

Initially, most of the heat needed to drive the endothermic RWGS reaction would probably come from combustion of lowest-value byproducts from the FTS reactions – primarily methane. Some of the heat may come from reforming of low-value FTS products (methane, ethane, and propane) into syngas using an exothermic partial oxidation. Concentrated solar heat could also often be used – even at night, with thermal storage.

Theoretically, it is not necessary for the RWGS temperature to be below the temperature of the FTS reactor to get a significant portion of the RWGS heat needed from the FTS reactor. A heat pump might be able to pump heat from the FTS reactor to a higher temperature with less electrical power than would be required for direct heating of RWGS reactors. However, effective heat pumps for this temperature range have not been shown to be practical.

With current catalysts the FTS production of light olefins, gasoline, or mid-alcohols can work with good selectivity, adequate lifetime, and acceptably low coking and methane production at least up to 600 K, making it easier to utilize its waste heat more efficiently. However, this is still not high enough to readily drive the RWGS reaction, at least under variable conditions – except perhaps if both the CO and the H_2O in the RWGS reactor are held to low levels. The amount of heat required for the RWGS is at least 8% that of the total FTS products. However, the difference in system efficiency between the two options (burning low-value byproducts or using FTS heat) is only about half that amount if highly effective methods are available for conversion of waste heat to electrical power. Still, the potential efficiency advantage of driving the RWGS with FTS waste heat provides incentive to develop catalysts and plant designs compatible with higher FTS temperatures and lower RWGS temperatures.

While the WGS is not a significant loss route in LT-FTS diesel or gasoline reactors, WGS activity in HT-FTS reactors can be quite high. Previously, there has not been a very good method for utilizing the WGS products and waste heat. However, when the syngas is being generated by the RWGS, the WGS byproducts can readily be converted back into syngas.

Approaching Second-law Limits in Waste Heat Conversion. There are two huge and comparable sources of waste heat in the WindFuels plant – the electrolyzer and the FTS reactor. Initially, these will probably be at just 10 K above current best practice in related applications (i.e., about 430 K and 600 K respectively), and each would be rejecting 30-60 MW in a 250 MW wind-fuels plant. Some of this can be used directly in preheating of reactants and in distillations of products, but most will need to be converted into electricity as efficiently as possible.

A large number of variations on the Organic Rankine Cycle (**ORC**) have been evaluated for the economic utilization of low-grade heat (as from geothermal sources, ~400 K) or mid-grade heat (as from concentrated solar power **CSP**, ~650 K). Few have exceeded 55% of the second-law efficiency limits. This is largely because the latent heat of vaporization and the differences in specific heats between the liquid and gas phases make full optimization (minimizing irreversibilities) impossible for a single heat source. We show in a pending patent application how a novel Dual-source doubly-recuperated Organic Rankine Cycle (**DORC**) allows one to achieve efficiencies much closer to second-law limits while simultaneously reducing the cost and complexity of the heat engine when both a low-grade and a mid-grade heat source (of comparable magnitudes) are available and the working fluid is optimally selected. The novel design is related to the dual-source steam Rankine cycle disclosed by Martin et al in US Pat 3,950,949. Simply put, two different heat sources are much better than one.

Designing for Variability. A widely noted characteristic of wind and solar energy is that they are variable. Wind and solar installations are customarily rated according to their peak power capability. Mean power generation in a typical Class 5 wind site is about 35% of the peak capability of the hardware. Mean power from CSP plants is typically only 18-26% of peak, though the CSP diurnal cycle is a usually a much better match to the grid demand cycle than is wind's.

We choose to identify RFTS plants by their ***mean*** power, as this allows for fairer comparisons to conventional power plants. However, the assumption is that the RFTS plant needs to be able to operate, essentially continuously, at three times this mean rating and perhaps at one-tenth, or at least at one-third, this mean rating. Geothermal, hydrokinetic, wave, and tidal energy are much less variable; and this would allow savings in an RFTS plant driven by such sources compared to one driven by wind or CSP.

The hydrogen production rate from the electrolyzers at the plant would be able to change as quickly as needed in response to changing wind conditions as long as the electrolyzer is maintained near optimum operating temperature and pressure. However, the RFTS plant would not be able to respond as quickly, so some local hydrogen storage would be needed – at least 6 hours worth – for efficient power-down, standby, and power-up cycles. For a plant of 250 MW ***average*** power, that comes to a fairly substantial amount – about 30 tons (360,000 m^3 at STP). While compact, light-weight hydrogen storage in small quantities (as needed for fuel-cell vehicles) is quite expensive, bulk hydrogen-gas storage at moderate pressure (1-15 MPa) is reasonable. Some carbon monoxide storage would improve transient response time and greatly simplify control during transients (as would the storage of some additional hydrogen at several different temperatures and pressures).

The huge amount of compressed oxygen byproduct being produced by the electrolyzers may saturate the local oxygen market, so it is useful to find a way to use some of it on site. Several ways that it may be advantageously utilized for improved plant efficiency are discussed in the pending patent. Sufficient amounts of compressed or liquid oxygen (**LOX**) would normally be stored for efficient refrigerator operation during power down, standby, and power up. The 250 MW plant, for example, would produce nearly 40 tons per hour. Such a plant may need to store more than 200 tons of LOX. These suggested H_2 and LOX minimum storage amounts are about one-third the fuel-up requirements for the space shuttle and are not difficult to accommodate safely. Extended hydrogen storage in quantities larger than ~50 tons may be better handled as a cryogenic liquid to accommodate days of excess wind capacity followed by light winds.

The plant site might be over 300 km from where most of the renewable energy or any other input originates, as distribution costs for all of the inputs and outputs (electricity, CO_2, water, liquid fuels, O_2, chemicals, waste heat, etc.) must be weighed.

Conclusions and Discussion

Simulations have been carried out in detail of a novel plant design for producing mostly ethanol from wind energy, waste CO_2, and water using well-validated, commercial software. The combination of 8 major innovations, detailed elsewhere, appears to promise system efficiency of 59% from wind energy to stable fuels in the near term. The cost of producing chemicals and fuels in the RFTS plant will depend mostly on the quality of the wind site and on the market for the co-produced liquid oxygen. In a Class-6 wind site, ethanol should be profitable at $1.20/gal as long as the local oxygen market is strong and subsidies for renewable fuels are not decreased. In a Class-4 site with no oxygen market and no subsidies for climate benefit, the cost of wind-ethanol should be about $2.90/gal. Wind's growth rate is currently beginning to be limited by transmission-grid capacity, but RFTS completely eliminates that problem.

The WindFuels concept builds on much of the progress made toward a hydrogen economy, but it recognizes the fundamental advantages of efficiently putting green hydrogen into stable liquid fuels and chemicals. The CO_2 for these products will initially come from the exhausts of power plants, as they currently provide an abundant supply. Ultimately, the CO_2 can come from the atmosphere. We believe a major, well focused development effort on WindFuels offers more promise for cost-effective, essentially limitless, carbon-neutral transportation fuels than any other avenue currently being explored.

References

1. AP Steynberg and ME Dry, eds. *Studies in Surface Science and Catalysis 152, Fischer-Tropsch Technology*, Elsevier, 2004.
2. Doty Energy website, http://windfuels.com/ , 2008.
3. S Phillips, A Aden, J Jechura, and D Dayton, "Thermochemical Ethanol via Indirect Gasification and Mixed Alcohol Synthesis of Lignocellulosic Biomass", NREL/TP-510-41168, 2007. http://www.nrel.gov/docs/fy07osti/41168.pdf
4. K Ibsen, "Equipment Design and Cost Estimation for Small Modular Biomass Systems. Task 9: Mixed Alcohols from Syngas – State of the Technology", NREL/SR-510-39947, 2006. http://www.nrel.gov/docs/fy06osti/39947.pdf
5. PL Spath and DC Dayton, "Preliminary Screening – Technical and Economic Assessment of Synthesis Gas to Fuels and Chemicals with Emphasis on the Potential for Biomass-Derived Syngas", http://www.fischer-tropsch.org/DOE/DOE_reports/510/510-34929/510-34929.pdf , NREL/TP-510-34929, 2003.
6. R Zubrin, B Frankie, and T Kito, "Mars In-Situ Resource Utilization Based on the Reverse Water Gas Shift: Experiments and Mission Applications", AIAA 97-2767, 1997, http://www.marssociety.de/downloads/Artikel/in-situ.pdf
7. M Xiang, D Li, H Qi, W Li, B Zhong, Y Sun, "Mixed alcohols synthesis from CO hydrogenation over K-promoted β-Mo_2C catalysts", *Fuel 86*, 1298-1303, 2007.
8. X Li, L Feng, Z Liu, B Zhong, DB Dadyburjor, and DL Kugler, "Higher Alcohols from Synthesis Gas Using Carbon-Supported Doped Molybdenum-Based Catalysts", *Ind. Engr. Chem. Res.* 37, 3853-3863, 1998.
9. K Weissermel, HJ Arpe, *Industrial Organic Chemistry*, 4th ed., Wiley, 2003.

10. Matthew Simmons, http://www.simmonsco-intl.com/research.aspx?Type=msspeeches , 2008.
11. Joe Romm, "The Self-limiting Future of Nuclear Power", Climate Progress, 2008, http://www.americanprogressaction.org/issues/2008/nuclear_power_report.html .
12. Adam Brandt, "Converting Green River oil shale to liquid fuels with ATP and ICP technologies: energy efficiency and GHG emissions", Univ. CA, Berkeley, 2007, **http://abrandt.berkeley.edu/shale/Brandt_Converting_Green_River_oil_shale_to_liquid_fuels.pdf**
13. G Olah and A Molar, *Hydrocarbon Chemistry*, 2nd ed., Wiley, 2003.
14. CH Bartholomew and RJ Farrauto, *Industrial Catalytic Processes*, Wiley, 2006.
15. JD Seader and EJ Henley, *Separation Process Principles*, 2nd ed., Wiley, 2006.
16. J Ivy, "Summary of Electrolytic Hydrogen Production", NREL/MP-560-36734, 2004, http://www.nfpa.org/assets/files/PDF/CodesStandards/HCGNRELElectrolytichydrogenproduction04-04.pdf
17. JI Levene, "Economic Analysis of Hydrogen Production from Wind", WindPower 2005, Denver, NREL/CP-560-38210, 2005. http://www.nrel.gov/docs/fy05osti/38210.pdf
18. R Bourgeois, "Advanced Alkaline Electrolysis", DE-FC36-04G014223, GE Global Research Center, 2006. http://www.hydrogen.energy.gov/pdfs/review06/pd_8_bourgeois.pdf .
19. DESIGN II for Windows Tutorial and Samples Version 9.4, 2007, by WinSim Inc., available from http://www.lulu.com/includes/download.php?fCID=390777&fMID=810115 .
20. JE Whitlow and C Parrish, "Operation, Modeling and Analysis of the Reverse Water Gas Shift Process", 2001 NASA/ASEE summer program, JFK Space Center, http://ntrs.nasa.gov/archive/nasa/casi.ntrs.nasa.gov/20020050609_2002079590.pdf
21. JF Decarolis, DW Keith, "The economics of large-scale wind power in a carbon constrained world", Policy, pp395-410, 2006.
22. JR Hensman, D Newton, "Fischer-Tropsch Synthesis Process", US Pat #7,115,670, 2006.
23. AP Steynberg, JW De Boer, HG Nel, WS Ernst, JJ Liebenberg, "Process for Synthesizing Hydrocarbons", Pending Pat. appl. pub. US 2007/0142481.
24. AJ Severinsky, "Systems, Methods, and Compositions for Production of Synthetic Hydrocarbon Compounds", Pending Pat. pub. US 2006/0211777.
25. JA Hogendoorn, WPM van Swaaij, GF Versteeg, "The absorption of carbon monoxide in COSORB solutions: absorption rate and capacity", Chem. Engr. J. 59, 243-253, 1995, http://doc.utwente.nl/11240/1/Hogendoorn95absorption.pdf
26. JF Martin and WL Kubic, "Green Freedom, A concept for Producing Carbon-Neutral Synthetic Fuels and Chemicals", Los Alamos National Laboratory, 2007, http://www.lanl.gov/news/newsbulletin/pdf/Green_Freedom_Overview.pdf
27. David G Wilson and Jon Ballou, "Design and Performance of a High-Temperature Regenerator Having Very High Effectiveness...", paper GT 2006-90096, Turbo-Expo 2006, Barcelona.
28. FG Kerry, *Industrial Gas Handbook – Gas Separation and Purification*, CRC Press, Boca Raton, 2007.
29. FD Doty, G Hosford, JB Spitzmesser, and JD Jones, "The Micro-Tube Strip Heat Exchanger", *Heat Transfer Engr.,* 12, 3, 31-41, 1991.
30. DR Hardy and T Coffey, "Process for Producing Synthetic Liquid Hydrocarbon Fuels", Pending Pat. appl. pub. US2005/0232833.
31. FD Doty, "Hydrocarbon and Alcohol Fuels from Variable, Renewable Energy At Very High Efficiency", pending patent application, PCT to publish 10/08.

330d Supercritcal Phase Fischer Tropsch Synthesis from Coal Derived Syngas

Jeremiah Benoit[1], **Kanchan Mondal**[2], and Tomasz Wiltowski[2]. (1) Mechanical Engineering and Energy Processes, Southern Illinois University, Carbondale, IL 62901, (2) Department of Mechanical Engineering and Energy Processes, Southern Illinois University, Carbondale, IL 62901

The paper presents the results from research on conducting FT synthesis in supercritical CO2 from syngas typically produced from coal gasification and using a Fe-Zn-K catalyst. The data show and enhancement in the hydrocarbon selectivity and reduction in the parasitic loss of carbon efficiency due to carbon dioxide formation along with significant improvement in the conversion rates and an increase the life of the catalyst. This was achieved by employing a unique reactor setup that can conduct gas phase or supercritical phase FT synthesis in both batch or flow modes. The use of the supercritical CO2 (ScCO2) inhibited both methane and CO2 selectivities while enhancing the rates of synthesis and altering the Anderson Schulz Flory distribution by favoring the desorption of ƒÑƒ{olefins formed on the catalyst surface. In addition, the use of supercritical CO2 was found to prolong the life of the catalyst presumably by removing the heat of reaction from the catalysts surface and solubilizing the waxes that tend to deposit of the surface. Although not within the scope of this paper, the products from such a reactor system can be easily separated without the need of an additional unit process simply by tuning the pressure and temperature. The product spectrum and the selectivites for the different products will be presented. The effect of process parameters such as temperature, pressure, CO2 partial pressure on the product spectrum will also be presented. The clear increase in CO conversion at H2:CO ratio of 1:1 in supercritical phase as compared to gas phase reaction, the decrease in CO2 and CH4 selectivity, overall liquid product distribution finally an increase in the life of the catalysts as compared to gas phase synthesis clearly show the potential of the use of supercritical medium (ScCO2) for coal to liquid fuels conversion especially since the reactions can be conducted in a two phase regime without losing the benefits of the 3-phase slurry reactor systems.

330e Catalytic Decarboxylation of Pyrolytic Oils Model Compound on Ceria-Zirconia Mixed Metal Oxide

Vissanu Meeyoo, Mahanakorn University of Technology, 51 Cheum-Sampan Road, Nong-Chok, Bangkok, 10530, Thailand and Pramoch Rangsunvigit, The Petroleum and Petrochemical College, Chulalongkorn University, 254 Soi Chula 12, Phyathai Rd., Patumwan, Bangkok, 10330, Thailand.

Text Not Available.

330f Economic and Environmental Analysis of Hydrogen Production Strategies

Wei Yuan, Susilpa Bommareddy, Norman E. Sammons Jr., and Mario R. Eden. Department of Chemical Engineering, Auburn University, 230 Ross Hall, Auburn University, AL 36849-5127

Experimental researchers in the Consortium for Fossil Fuel Science (CFFS) are developing novel technologies for hydrogen production and storage, however the economic and environmental viability of the technologies are mostly evaluated separately rather than in an integrated manner. In this work, we aim to address this gap through the development of integrated system engineering strategies for modeling, integration and optimization of hydrogen polygeneration plants.

We are developing process models for a variety of hydrogen production schemes. Using process integration techniques and advanced computer-aided tools, the systems have been optimized and the economic potential as well as the environmental impact of the technologies evaluated. Several reforming techniques including four developed within CFFS have been studied based on small and large scale production of hydrogen. Literature data along with data obtained in other research projects at Auburn University has been used to develop models for partial oxidation (POX), steam (SR) and autothermal reforming (ATR) of a variety of hydrocarbon resources. In addition, data provided by other researchers in CFFS has been used to build similar models for super critical water reforming (SCWR) of methanol and biomass (using glucose as a model molecule), dry reforming (DR) of methane and finally a catalytic dehydrogenation (CDH) of methane. The last process is a single step process that in addition to high purity hydrogen also produces a valuable carbon nanotube byproduct.

Our work has made it clear, that technologies like POX and ATR due to their exothermic nature are much less sensitive to changes in heating utility cost like the recent spike in natural gas prices. However it is worth noting that all processes can be improved immensely by implementing process integration. An economic analysis of all the generated case studies was performed to evaluate the hydrogen production cost. It was found that the current technologies for producing hydrogen from liquid fuels are not attractive if evaluated only on the production cost. Benefits such as transportability etc. will need to be quantified for all types of fuel in order to better compare the technologies. The environmental impact for each process is evaluated and compared. The results show that SCWR of biomass is most environmental friendly compared to the other processes. Several technologies as being developed by CFFS researchers appear to be potentially cheaper alternatives to the current state of the art.

330g Experimental Investigation of Methane Gas Production from Methane Hydrate

Yue Zhou[1], Marco J. Castaldi[2], and Tuncel Yegulalp[2]. (1) Earth and Environmental Department, Columbia University, New York, NY 10025, (2) Earth & Environmental Engineering (HKSM), Columbia Univeristy, 500 West 120th street, 926 Mudd Building, New York, NY 10027

Experimental investigation of methane gas production from methane hydrate

Abstract

Natural gas hydrates are solid, non-stoichiometric compounds of small gas molecules and water. They form when methane from organic decomposition comes together with water at low temperatures and high pressures to trap individual gas molecules within atomic scale crystalline cages of water ice. Natural gas hydrates have the potential to meet global energy needs for the foreseeable future. The total amount of methane gas which is entrapped in hydrates is more than twice the known fossil fuel. Intensified research worldwide is attempting to find a practical, safe, and environmentally acceptable approach to produce for the production of methane gas.

To test the concept of methane production via in-situ heating [1] and characterize other alternative production schemes, a reactor vessel has been designed, manufactured and built to test the concept model. Preliminary tests show that the reactor satisfies the conditions for methane hydrate and carbon dioxide formation. The thermocouples in the reactor detected the temperature change associated with methane hydrate formation. Hydrate was formed at 2.2 oC and about 600 psi with a concentration of 99.97% pure methane gas and DI water. Based on the mass balance calculation, 4.45% of hydrate was formed in the reactor at these conditions. Measurement and simulation suggest that hydrate was initially formed at the top section of the reactor followed by the hydrate formation within the sediment. During the dissociation via depressurization, a cooling effect from the endothermic reaction was observed. Also within a narrow range of temperature and pressure, a transition regime was observed during the hydrate dissociation test in which a large increase of gas production was recorded. It is first time reported as an experiment result, since Tsypkin's [2] predict the phenomenon in his calculation.

330h Sds Adsorption on Cyclopentane Hydrates at the Hydrate/liquid Interfaace

Chi Lo, Chem Eng Dept, The City College of New York, 140th St and Convent Ave, New York, NY 10031

Text Not Available.

Simulation and optimization of the process used in Colombia for the production of Biodiesel from palm oil: a kinetic analysis and an economical approach

F. Velosa[‡], J.M. Gomez[†‡].

Los Andes University, Carrera 1 N° 18A 10 Bogotá, (Colombia)
[‡]Chemical Engineering Department

Colombia, as a developing agricultural country interested on the progress of the biofuel industry, has acknowledged the oil market juncture, creating a legal frame to become one of the main biofuel players in the region. As palm growers and the industry invest on the developing business, the academy develops technical and economical tools that allow a promising start.

A simulation and optimization of the process used in Colombia for the production of Biodiesel from palm oil under homogenous alkali catalysis was developed. Biodiesel production was chosen as the objective function, obtaining a purity of 99.8 wt% according to the specifications established, as minimum conditions, by the ASTM D 6751 standard (99.6 wt%). Biodiesel and glycerin yields, as well as the water flow rate were found to be the most significant factors affecting the objective function. The production increment, compared with the base case (kg/hr) using numeric optimization was 5%, and the concentration of the glycerin as a result of the simulation, was high enough to sell it as a commodity product (99.4 wt%), accomplishing a more profitable production process.

Key Words: Palm oil, homogenous alkali catalysis, simulation and optimization.

1 Introduction

There are many cases throughout history that have shown the special value of using vegetable or waste oils in order to produce a renewable diesel fuel.

The first example came from a Belgian patent written by G. Chavane on August 31, 1937, which describes the use and production of palm oil ethyl esters as diesel fuel. During the 1930's and 1940's vegetable oils were used as diesel fuels from time to time, but usually only on emergency situations.

Another great example was when the United States developed a great concern, in the years after World War II, about the rising use of petroleum fuels and the possibility of resultant fuel shortages, developing what at the time was called a dual fuel project, involving institutions as The Ohio State University and Georgia Institute of Technology [1, 2].

At present, the instability and high oil prices, the decreased sources of fossil oil and the implementation of Kyoto Protocol to limit the greenhouse emissions, have been an inspiration to focus on vegetable oils and animal fats to produce biofuels [3]. The environmental aspect has received more attention since the clean developed mechanism was introduced on the global change negotiation in Kyoto. This is only a mechanism by which industrialized countries rely on the developing ones for contribution, accomplishing the weather stability [4].

Colombia, as a developing agricultural country interested on the progress of biofuel industry, has acknowledged this unique opportunity, and through the National Government has developed and implemented laws that regulate these sources of energy. The law 939 (December 30th, 2004) has established that diesel fuel, by August of the present year, has to be B05 which means that the content of biodiesel will be 5%. In other words, the demand conditions and the selling price are going to be determined by the legislation, establishing biofuels as a key element on the future of the country.

Recently, the world has shown a special interest on alternative energy sources, related to the imminent possibility of petroleum extinction; one of the oil substitutes analyzed is biodiesel, an important player on the Colom-

[†] To whom correspondence should be addressed. Tel.: +571 3393939 Ext.: 3744. Fax: +571 3324334. e–mail: jorgomez@uniandes.edu.co.

bian territory since this country has experienced in the past years an increase on diesel fuel consumption. During 2001, Colombia used 59,915 barrels/day; meanwhile, throughout 2005 consumed 81,800 barrels/day, experiencing a 36.5% growth in four years [5].

In addition, the palm oil production during 2001 was 10.718 barrels/day, from which 27% was destined to export, mainly to Europe, 68% was used for national consumption [6] and the additional 5% was available for the development of the biodiesel industry. Therefore, if the legislation had been implemented during 2001, the available palm oil could have only covered 18% of the demand of biodiesel, and during 2005, this value would have decreased to 10% [7]. Therefore, Colombia has a long way to go in order to be able to accomplish the projections stipulated for the third trimester of this year, and it will need an enormous support by the community and the government to achieve this process objective.

Considering the technology implemented in this country, the overall Colombian panorama results in a very interesting challenge for the national industry, as well as an opportunity to develop a solid case of study focused on the simulation and optimization of a biodiesel production mill using palm oil as raw material. At this specific point, it is important to mention that even though production industry is very recent, it is essential to employ computational tools, as the commercial software Aspen Plus®, since this software offers the possibility to: analyze the viability of the project in order to know the optimum operation conditions, evaluate particular cases of study, augment the knowledge of the process, and it could also be used as an alternative tool for training, as well as many other benefits that investors, engineers and workers can have.

Most of the recent studies about the Biodiesel production using basic catalysts are focused either on specific features of one of the individual processes necessary to develop a production mill, or they have been approaching the generalities and theoretical description of the global production system. Zhang et. al. (2003), as the base case of study, have done a technical and economical evaluation of the Biodiesel production process using basic (NaOH) as well as acid (H_2SO_4) catalysts. This research was approached by means of the simulation practice, using the commercial software HYSYS® Plant Net developed by Hyprotech Ltd., with the purpose of designing a complete and continuous process of the plant. The economic assessment took place simultaneously, identifying the aspects that had a strong influence on the financial viability of Fatty Acid Methyl Esters (FAME) production process.

Zhang et. al. (2003) presented an evaluation based in four different continuous process flowsheets for biodiesel production from virgin vegetable oil or waste cooking oil under alkaline or acidic conditions on a commercial scale. This case of study was divided in two; the fist part presented a comparison of these four processes from a technological point of view. The results demonstrate that the production of FAME from waste cooking oil, by means of acidic catalyst is a competitive alternative to be considered. Furthermore, in the second part of this work an economic evaluation of the viability of those four processes was done, identifying the most representative financial factors.

In a previous study three different simulations were conducted to identify the operational conditions that adapt to the needs of the market and the circumstances of the environment (Velosa et. al., 2007). In the next paragraphs a discussion will be effectuated around the kinetic analysis and economical factors surrounding biodiesel production. In order to improve the simulation and optimization of the process used in Colombia for the production of biodiesel from palm oil.

The relationship between the technical concepts and the economical factors is decisive when a project is evaluated, since those two perspectives are directly involved on its viability. A project's economical evaluation is a great complement to a technical approach, since it reduces the limitations between the academic exercise and reality, as well as allows a starting point for the election of the variables to be optimized.

The present article focuses on the biodiesel production optimization using an analysis of the process in order to determine the variables that assure the efficiency of the production mill, in addition to a rigorous analysis of transesterification kinetics. Furthermore, a cost estimate has been presented as an evaluation of the fixed investment cost necessary for the construction of a biodiesel production plant, in order to add an economical point of view to the technical approach that this study has to offer. It is important to mention, that along with all the advantages that the simulation software includes, it also has limitations related with its ability to model reality.

2 Background

2.1 Economic Studies

In the past few years, a number of papers have been written developing the economical cost analysis of biodiesel production mills. These publications focus on the economic assesment to determine the process viability and the advantages that each technology has over others [9, 11, 12].

Currently, the high cost of Biodiesel compared to the diesel fuel obtained by mineral sources, is one of the main impediments for its commercialization. It costs near 1.5 times more than the combustible based on petroleum. Between 60 and 75% of the cost of production is related to the cost of raw material [2, 8, 11]. While in Colombia the industry of palm oil finds a position on the global market and tries to generate a positive commercial balance as an export; the prices of growth, harvest, transportation and extraction have been above those of the main producers of palm oil worldwide, Malaysia and Indonesia [6].

Among the examined plant system variables: plant capacity, feedstock price, glycerin and biodiesel yields were found to be the most significant variables affecting the economic viability of Biodiesel manufacture. It is important to mention that these studies were also based on the work developed by Zhang et. al. (2003).

2.2 Trygliceride Transesterification

In general, this reaction takes place with a basic or acid catalyst. Basic catalysts are the most commonly used, mainly because the reaction is faster and the operation conditions are moderate [13].

$$T + 3ROH \rightarrow 3R'CO_2 - R + G$$

Figure 1 Overall Scheme of the Triglyceride Transesterification.

$$T + ROH \underset{k4}{\overset{k1}{\leftrightarrow}} D + R'CO_2 - R$$

$$D + ROH \underset{k5}{\overset{k2}{\leftrightarrow}} M + R'CO_2 - R$$

$$M + ROH \underset{k6}{\overset{k3}{\leftrightarrow}} G + R'CO_2 - R$$

Figure 2 Triglyceride Transesterification Reaction Scheme.

Transesterification, also called alcoholysis, involves the substitution of the alkyl group of the ester by another through the interaction between the ester and the alcohol. In other words, one of the carbon chains of the triglyceride is substituted by the OH group of the alcohol that reacts with it, to form the ester (Biodiesel). However, if the selected alcohol is methanol, the reaction could be called methanolysis. The vegetable oil methanolysis yields fatty acid methyl esters and glycerol; the first ones are excellent substitutes of diesel fuel [14].

Among the alcohols that could be used in the transesterification process it is typical to find methanol, ethanol, propanol and butanol. Methanol and ethanol are the most frequently used alcohols, especially methanol for its cost and its physical and chemical advantages as: polarity and shortest chain, being able to react in short periods of time with triglycerides and NaOH [2].

Moreover, most of the ethanol produced in Colombia has been assigned to the biofuel industry. It is blended with regular fuel in order to diminish the dependence on oil derived fuels among some other reasons.

Methanolysis would proceed at standard temperature (21 ºC) with a reaction time of 4 to 8 hours. The reaction time could be diminished to 1 or 2 hours rising the temperature up to 60 ºC [12, 15, 16]. Lower operation temperatures suggest lower energy consumption and a decrease in investment costs, but the raw material has to have a higher purity [17].

Even though this kind of system holds great benefits in terms of time of reaction, it has a very particular limitation which is the purity of the raw material. The production of esters by means of basic accelerant agents is sensitive to the water as well as the free fatty acid content. The presence of water may cause the ester saponification under basic conditions. In contrast, the free fatty acids could react with the catalyst getting soaps and water as intermediate products. The saponification not only consumes the catalyst but causes the production of emulsions which generates problems in FAME recovery and purification [2, 9, 14].

The pretreatment of the raw material is a very important operation since it diminishes the water and free fatty acid content to less than 0,1 and 0,05 wt%, respectively, in order to obtain a high degree of purity.

The triglyceride transesterification reaction could be catalyzed by sodium hydroxide or potassium hydroxide dissolved on methanol. Even though these two types of catalyst have being used frequently among the researches through time, this precise case of study has chosen NaOH as its catalyst.

$$NaOH + CH_3OH \rightarrow NaOCH_3 + H_2O$$

Figure 3 Reaction Scheme for the Catalyst Activation (NaOH).

Figure 3 illustrates the catalyst activation reaction, where the active catalyst structure is represented by the alkoxide which is formed in the methanolic solution [17, 18].

3 Process Design

Simulation is one of the tools used by chemical engineers to interpret flowsheet diagrams, identify problems and predict the process performance. The core of the analysis is the mathematical model, a set of equations that relate the process variables. Steady state simulators as Aspen Plus® solve the unknown variables starting from specified quantities.

There are several levels of complexity involved on a project evaluation: mass and energy balances (short calculation), equipment sizing (rigorous calculation) and profitability analysis [19].

The development of this case of study was done by simulating a process using a short and rigorous calculation in order to determine the dimensions and characterization of the equipment. As a result, the objective function was built choosing the quantity of methyl oleate (biodiesel) as the variable to be optimized. It is important to mention that two cases of study were developed based on the article *Biodiesel production from waste cooking oil: 1 Process design and technological assessment;* written by Zhang et. al. (2003).

The main plant units include CSTR reactors, distillation columns and extraction columns. The materials required for the main process units are were obtained from the base case of study. Taking into account the corrosive character of sodium hydroxide, the suggested material for the reactors is stainless steal 316; for the rest of the units, carbon steel will be used [9].

Distillation is chosen as the separation process for methanol, as well as for the purification of FAME and glycerol. Even though the boiling point of methanol (65 °C at 1 atm) is lower than FAME's (approximately 320 °C at 1 atm) or glycerol (300 °C at 1 atm), simulations have suggested that the desired purities for biodiesel and glycerol (grater that 90 wt%) can not be achieved by a flash unit [10]. For this type of unit, it is assumed that the tray efficiency was 60%, in addition to a security factor of 15%.

3.1 Reactors (R-101 y R-101A)

Many studies of alkali catalyzed transesterification based on simulations have been elaborated; most investigated the economic feasibilities of the different technologies used to produce Biodiesel. For these articles, a conversion reactor was chosen in spite of the availability of kinetic data, for all of the different types of catalysts, in order to be able to compare the results of the selected technologies [8, 9, 10, 11, 12].

For this particular case of study, the reaction and its parameters were included, in the rigorous calculation and the optimization, to formulate a scenario that reduces to a certain point, the limitations related with the ability of the software to model reality.

The reaction was carried out at a 7: 1 molar ratio of alcohol to palm oil, since at high molar ratios the ester conversion increases as the reaction evolves faster [2]; a 1% weight catalyst concentration, which was based on oil, (basic catalysis). This specific concentration has been justified by studies that Darnoko et. al. (2000) have made where they acknowledged that it was the optimum concentration.

The operation conditions for the reactor were fixed at 80 °C and 400 kPa, differing from the 60 °C temperature mention by Zhang et. al. (2003). The reactor pressure was established considering that methanol boiling point (65 °C a 1 atm) is an important constrain. Moreover, many studies of alkali-catalized transesterification have been carried out, in which a reaction temperature near the alcohol's boiling point was recommended [8, 9, 11, 14, 16]. It is critical to consider that at values bellow 100 °C the formation of soaps occurs if the purity of the raw material is not evaluated as an extremely important process variable [17]. Different Industrial approaches have shown that the values for this variable vary form 70 to 80 °C [20].

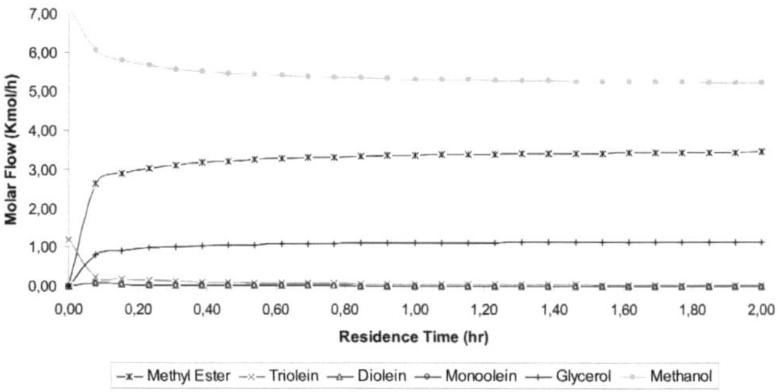

Figure 7 Flow Profile against Residence Time R-101.

Another important variable is the residence time which could be determined by means of a flow profile analysis. Figure 7 illustrates the reactants' and the products' flow profile against residence time. The results from a kinetic point of view; have a very satisfactory tendency compared with the experimentation proposed by Gemma et. al. (2005); residence time was 1.5 hours on each one of the reactors and the volume was 2.4 m^3. Since an increase on the time variable would not affect the production considerably.

A series pattern was chosen for this particular case because it increases the conversion form 96.1% to 99.8%, which facilitates the separation and purification processes, making the plant more efficient. Being one of the most significant changes effectuated compared with the base case. It is important to mention that this modification represents only a 9% increase in the total capital investment.

The composition obtained, based on palm oil, for the first reactor was of 34.7% and for the second reactor it increased to 36.0%, value that reflects the importance of the improved pattern.

The reactors were assumed to operate continuously for all cases, Lab-scale conditions and reaction analysis made by Gemma et. al. (2005) for rigorous calculation and optimization were assumed to be appropriate for large scale production. The kinetics of triolein reactions are based on a reaction mechanism that includes a mass transfer-controlled region preceded by a kinetic section. Nevertheless, the mass transfer controlling segment is insignificant while using an impeller speed of 600 rpm. The result for the experiment demonstrates that the kinetic section follows a second order mechanism for the reactions, where the reaction system can be presented as a pseudo homogeneous catalyzed reaction [14]. Thus, the equations regarding the previous mentioned system are as follows:

$$\frac{dT}{dt} = -(k_1C)[T][A] + (k_2C)[E][D] \quad (1)$$

$$\frac{dD}{dt} = (k_1C)[T][A] - (k_2C)[E][D] - (k_3C)[D][A] + (k_4C)[E][M] \quad (2)$$

$$\frac{dM}{dt} = (k_3C)[D][A] - (k_4C)[E][M] - (k_5C)[M][A] + (k_6C)[E][G] \quad (3)$$

$$\frac{dE}{dt} = (k_1C)[T][A] - (k_2C)[E][D] + (k_3C)[D][A] - (k_4C)[E][M] + (k_5C)[M][A] - (k_6C)[E][G] \quad (4)$$

$$\frac{dA}{dt} = -(k_1C)[T][A] + (k_2C)[E][D] - (k_3C)[D][A] + (k_4C)[E][M] - (k_5C)[M][A] + (k_6C)[E][G] \quad (5)$$

A reaction analysis reveals that the values of the k_6C constants, corresponding to the reaction of the glycerin with the methyl ester to give monoglyceríde and methanol were not significant. In other words, the reaction was not favored mainly because of the immiscibility of methyl esters and glycerol, which involved a great mass transfer resistance in that direction [14, 16].

An alternative for low methanol consumption is the implementation of a simultaneous reaction and separation system for the formed glycerol, which could be done by a selective membrane [17]. Unfortunately, when the glycerin is removed a big quantity of catalyst will be also removed; therefore, it is necessary to add extra catalyst and methanol to the second CSTR reactor [18]. This type of equilibrium based system is used by the national industry in order to diminish costs since there is no methanol production in the country.

It is important to evaluate the financial effect of this type of upgrades since it is necessary to invest in two membranes, instead of one methanol recovery system. In other words, it is appropriate to balance the economical implications that have two membrane reactors instead of a low methanol flow on stream 101. A lower capital cost could be represented by the system proposed on this article, suggesting a multi-stage column for methanol separation, in order to incorporate it into the reaction system.

3.2 Water washing (T-301)

The purpose of this step is to separate FAME from glycerol, methanol and the catalyst, by means of a washing column with four equivalent theoretical stages. It is essential to point out that this specific unit uses the UNIFAC-Dortmund thermodynamical model, introduced by Weidlich and Gmehling in order to overcome the disadvantages of the UNIFAC model, such as the short temperature range (10 - 40 ºC) and the unsatisfactory result for the activity coefficient prediction; for those systems where different sized molecules are involved. In addition, the predictions for the behavior of systems that have alcohol are in some opportunities also incorrect [9, 21].

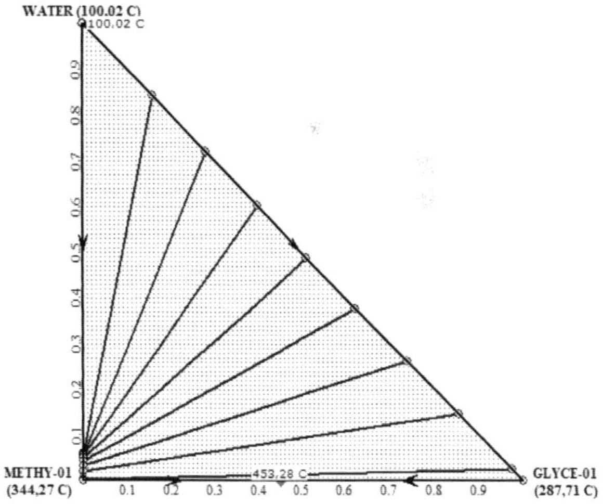

Figure 6 Ternary Map for the Washing Column T-301.

For this type of operations, it is recommended to use soft water to avoid the potential mineral transfer (calcium, magnesium and iron) to biodiesel. The temperature recommended to obtain satisfactory values of glycerin removal is 60 ºC [18].

The previous work did not incorporate sodium hydroxide, as one of the raw material stream, because the parameters established by the UNIFAC-Dortmund thermodynamical model for this reactant were not available. For this work a more advanced Aspen Plus® version was used in order to simulate the production plant including the catalyst, from the beginning of the process.

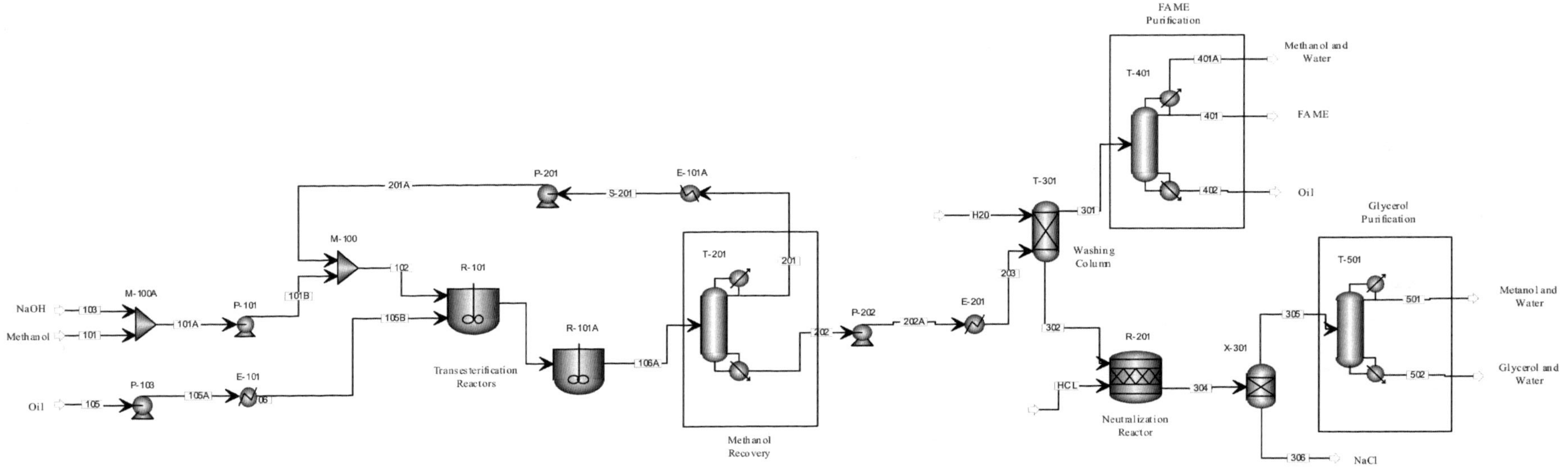

Streams	*101*	*102*	*105*	*106A*	*201*	*202*	*203*	*301*	*302*	*304*	*305*	*306*	*401*	*401A*	*402*	*501*	*502*
Temperature C	25,00	24,23	50,00	80,00	28,34	78,54	60,00	57,96	58,63	60,00	60,00	60,00	193,70	193,70	257,56	50,53	136,91
Pressure atm	0,99	3,95	0,99	3,95	0,20	0,20	1,97	1,09	1,09	1,09	1,09	1,09	0,10	0,10	0,10	0,49	0,49
Vapor Fraction	0,00	0,00	0,00	0,00	0,00	0,00	0,00	0,00	0,00	0,00	0,00	0,01	0,00	1,00	0,00	0,00	0,00
Mole Flow kmol/hr	5,20	8,67	1,19	9,86	3,47	6,39	6,39	4,05	2,94	3,44	3,19	0,26	2,85	0,53	0,67	1,67	1,52
Mass Flow kg/hr	166,62	277,81	1050,00	1327,81	111,19	1216,62	1216,62	1067,91	159,71	178,83	164,11	14,72	842,22	26,45	199,24	48,92	115,19
Volume Flow l/min	3,48	5,79	66,64	26,42	2,33	24,09	23,71	21,06	2,41	2,52	2,49	1,06	18,77	3452,80	4,79	1,04	1,61
Enthalpy MMBtu/hr	-1,18	-1,96	-2,06	-4,21	-0,78	-3,45	-3,49	-2,50	-1,15	-1,30	-1,22	-0,08	-1,67	-0,12	-0,36	-0,39	-0,81
Component Mass Fraction																	
Methanol	1,000	1,000	0,000	0,124	1,000	0,043	0,043	0,009	0,272	0,243	0,265	0,000	0,000	0,351	0,000	0,881	0,003
Triolein	0,000	0,000	1,000	0,001	0,000	0,001	0,001	0,002	0,000	0,000	0,000	0,000	0,001	0,001	0,006	0,000	0,000
FAME	0,000	0,000	0,000	0,793	0,000	0,865	0,865	0,986	0,000	0,000	0,000	0,000	0,999	0,513	0,993	0,000	0,000
Diolein	0,000	0,000	0,000	0,000	0,000	0,000	0,000	0,000	0,000	0,000	0,000	0,000	0,000	0,000	0,001	0,000	0,000
Monoolein	0,000	0,000	0,000	0,000	0,000	0,000	0,000	0,000	0,000	0,000	0,000	0,000	0,000	0,000	0,000	0,000	0,000
Glycerol	0,000	0,000	0,000	0,082	0,000	0,090	0,090	0,000	0,682	0,609	0,664	0,001	0,000	0,000	0,000	0,000	0,946
Sodium Hidroxide	0,000	0,000	0,000	0,000	0,000	0,000	0,000	0,000	0,000	0,001	0,000	0,014	0,000	0,000	0,000	0,000	0,000
Water	0,000	0,000	0,000	0,000	0,000	0,000	0,000	0,003	0,046	0,065	0,071	0,000	0,000	0,135	0,000	0,119	0,051
Hydrogen Chloride	0,000	0,000	0,000	0,000	0,000	0,000	0,000	0,000	0,000	0,001	0,000	0,012	0,000	0,000	0,000	0,000	0,000
Sodium Chloride	0,000	0,000	0,000	0,000	0,000	0,000	0,000	0,000	0,000	0,080	0,000	0,973	0,000	0,000	0,000	0,000	0,000

Figure 4 Flowsheets of the Process used in Colombia for the Production of Biodiesel from Palm Oil (Rigorous Calculation).

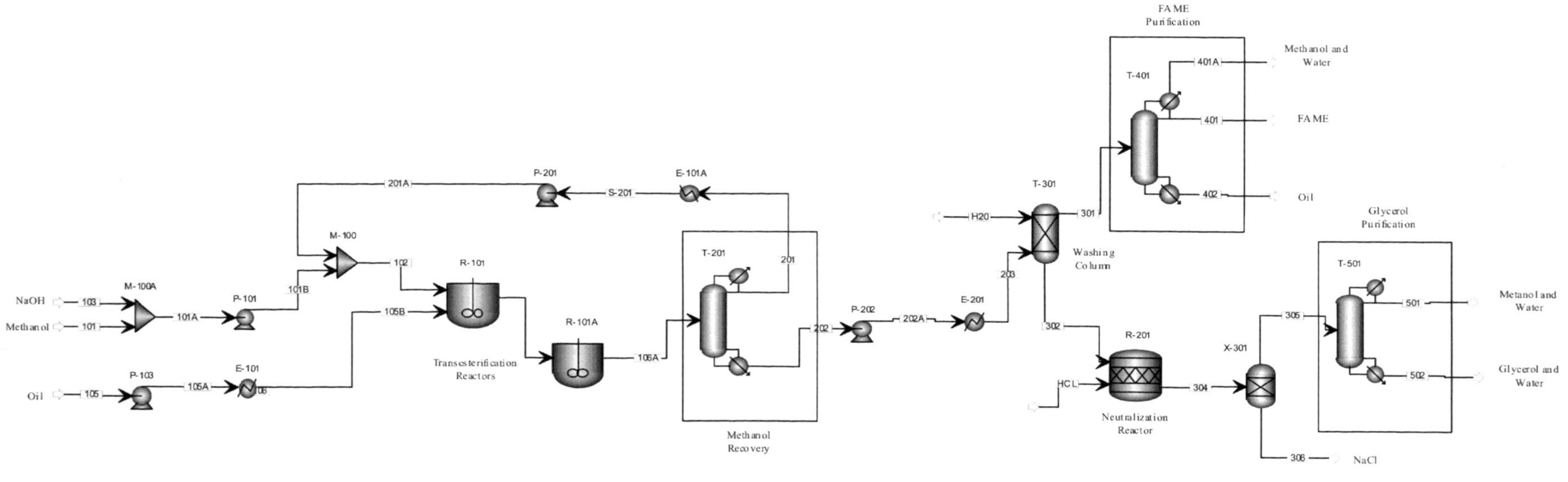

Streams	101	102	105	106A	201	202	203	301	302	304	305	306	401	401A	402	501	502
Temperature C	25,00	23,34	50,00	80,00	33,30	262,28	60,00	58,10	59,38	60,00	60,00	60,00	193,70	193,70	619,48	28,86	74,92
Pressure atm	0,99	3,95	0,99	3,95	0,20	0,20	1,97	1,09	1,09	1,09	1,09	1,09	0,10	0,10	0,10	0,04	0,04
Vapor Fraction	0,00	0,00	0,00	0,00	0,00	0,00	0,00	0,00	0,00	0,00	0,00	0,05	0,00	1,00	0,00	0,00	0,00
Mole Flow kmol/hr	3,43	7,15	1,19	8,34	3,47	4,87	4,87	3,71	1,77	2,02	1,77	0,25	3,48	0,21	0,01	0,38	1,38
Mass Flow kg/hr	110,00	283,08	1050,00	1333,08	163,08	1170,00	1170,00	1059,21	121,72	130,84	116,40	14,44	1048,62	9,08	1,50	6,88	109,52
Volume Flow l/min	2,30	5,09	66,64	26,96	2,69	28,80	23,98	21,53	1,77	1,64	1,55	5,97	23,95	1369,97	0,04	0,12	1,46
Enthalpy MMBtu/hr	-0,78	-1,97	-2,06	-4,22	-1,13	-2,63	-3,16	-2,44	-0,88	-0,96	-0,88	-0,08	-2,07	-0,05	0,00	-0,10	-0,78
Component Mass Fraction																	
Methanol	1,000	0,685	0,000	0,063	0,515	0,000	0,000	0,000	0,000	0,000	0,000	0,000	0,000	0,000	0,000	0,000	0,000
Triolein	0,000	0,000	1,000	0,026	0,000	0,029	0,029	0,032	0,000	0,000	0,000	0,000	0,032	0,032	0,011	0,000	0,000
FAME	0,000	0,005	0,000	0,765	0,009	0,870	0,870	0,961	0,000	0,000	0,000	0,000	0,966	0,584	0,000	0,000	0,000
Diolein	0,000	0,000	0,000	0,002	0,000	0,002	0,002	0,002	0,000	0,000	0,000	0,000	0,001	0,000	0,768	0,000	0,000
Monoolein	0,000	0,000	0,000	0,000	0,000	0,000	0,000	0,001	0,000	0,000	0,000	0,000	0,001	0,000	0,000	0,000	0,000
Glycerol	0,000	0,274	0,000	0,137	0,476	0,090	0,090	0,000	0,864	0,804	0,903	0,000	0,000	0,000	0,000	0,000	0,960
Sodium Hidroxide	0,000	0,035	0,000	0,008	0,000	0,009	0,009	0,000	0,078	0,001	0,000	0,013	0,000	0,000	0,221	0,000	0,000
Water	0,000	0,000	0,000	0,000	0,000	0,000	0,000	0,004	0,058	0,086	0,096	0,000	0,000	0,384	0,000	1,000	0,040
Hydrogen Chloride	0,000	0,000	0,000	0,000	0,000	0,000	0,000	0,000	0,000	0,005	0,000	0,042	0,000	0,000	0,000	0,000	0,000
Sodium Chloride	0,000	0,000	0,000	0,000	0,000	0,000	0,000	0,000	0,000	0,104	0,000	0,945	0,000	0,000	0,000	0,000	0,000

Figure 5 Flowsheets of the Process used in Colombia for the Production of Biodiesel from Palm Oil (Optimization).

4 Process Simulation

The Colombian Biodiesel demand and the acquisition of European technology that has not been proved in the country, generates an uncertainty based on the lack of information and experience on the production of alternative sources of energy, especially biodiesel. Nevertheless, this uncertainty could be minimized implementing engineering software to simulate and optimize the process, accomplishing an important technical development, in order to be one step ahead of an industry that represents the future of alternative fuels.

Despite the differences between the academic exercise and reality, the simulation of the biodiesel production process using sodium hydroxide as catalyst was performed employing the process simulation software Aspen Plus®., because of its complete compound library, thermodynamic packages, as well as its industrial worldwide recognition.

The process simulation procedures involved essentially: the definition of the compounds, selection of the thermodynamic model, determination of plant capacity, units and operation conditions.

The plant capacity was defined as one of the main parameters in order to establish the comparative procedure necessary to develop this work. This article chose the palm oil inlet 1050 kg/hr as the reference point; because this would allow the authors to define the production as the variable to be optimized.

Palm oil was chosen as raw material since it represents one of the most important sources of vegetable oil of the Colombian industry. In this particular case the molecule implemented in the simulation process was Triolein ($C_{57}H_{104}O_6$), given that it constitutes 40.5% of palm oil. Moreover, the catalyst was sodium hydroxide and the alcohol was methanol. Consequently, Methyl Oleate ($C_{19}H_{36}O_2$) was the product of interest. Properties and structure were factors established in the software library for the majority of the compounds, in order to start up the simulation. For compound like: Triolein (T), Diolein (D) and Monolein (M), this type of parameters were not available; for this reason it was necessary the implementation of the software ChemDraw Ultra®, in order to import the structure by means of the tool *User Defined* of Aspen Plus® components menu. Using the group contribution method, the software internally calculates the properties needed to simulate this case of study.

Carlson (1996) from Aspen Technology Inc. suggests the use of NRTL (Non Random Two Liquid) and UNIQUAC (Universal Quasi-Chemical) to predict the component activity coefficients on the liquid phase, based on the presence of highly polar compounds as water, methanol and glycerin, components with medium polarity as diolein and monolein, as well as non polar compounds as triolein and methyl oleate. Seeking for the flexibility of the case of study, NRTL was established as the main thermodynamical model. In addition, UNIQUAC was used as a support model through the tool *Referenced* located in the property menu, this was supported on three reasons: 1) UNIQUAC has two adjustable parameters while NRTL has three, 2) UNIQUAC parameters have a lower sensitivity to temperature and finally, 3) since the dependent variable of concentration is a superficial fraction and not a molar fraction, UNIQUAC is applicable to solutions that have small and big molecules [23]. In conclusion, having two complementary thermodinamical models allows the simulation to have a balance between the complexity of those models and the reliability of the results.

5 Optimization Problem Description

The optimum value is the solution that offers the most efficient and enhanced benefit to cost ratio in a problem or process design. Optimization is one of the most quantitative tools in the industrial decision making process.

The purpose of optimization is to find the value of the variables involved in the process that produced the best performance criteria (objective function).

It is important to systematically identify the objective function, constrains and degrees of freedom in a plant, since this will benefit and improve the design, obtaining a fast and reliable solution to problems, and an opportune as well as accurate decision making [24].

The relationship between the optimization and the economical factors is decisive when a complete analysis is preformed, since those two main approaches affect directly the viability of a project.

Economic considerations are the key driving force supporting the development of inexpensive production processes. Although, Biodiesel production total cost depends heavily upon feedstock costs, there are technological parameters that affect the plant's total manufacturing costs.

5.1.1 Objective function and constrains

The objective function for this case of study has been defined as the production of biodiesel, to identify the operational conditions which make the process as efficient as possible, and since it was identified as one of the most significant variables by the economical studies mentioned.

Furthermore, constrains were separated in two groups: process constrains and design constrains. Process constrains were stipulated by the ASTM D 6751 standard, specifying the biodiesel quality regulation. Design constrains were specified by the equations of each one of the units that structure the production plant.

5.1.2 Degrees of freedom analysis

After the simulation process has been accomplished all the equipment variables involved in the models are calculated, although only a few of them can be manipulated by the personal that closely interact with the process [24].

As the Colombian government decided to encourage national energy sources such as biofuels for the transportation sector, and since the viability of this projects rely on the state intervention; the economical factor has to be analyzed in order to become successful in the future. In other words, as the national subsidies are diminishing, the market will start to open, and only the economical efficient plants are going to prosper.

Among the system variables of the plants examined by most of the economical publications mentioned through this article, plant capacity, price of feedstock, and glycerin and biodiesel yields, were found to be the most significant variables affecting the economic viability of biodiesel manufacture. Therefore, since plant capacity is a fixed parameter and feedstock price depends directly on the market, the distillate flow rate for the FAME production tower, the extraction solvent flow and the methanol flow rate are the controllable variables that this work is focused on.

5.1.3 Sequential Quadratic Programming (SQP)

This type of numerical solution method solves a programming sequence of quadratic approximations, as to resolve a non linear programming problem. The method is based on a recurrent solution which iterates by means of Newton's algorithm, with the purpose of finding the optimal solution, applying the first order necessary conditions for inequality restriction problems; this means the use of necessary as well as sufficient conditions (Kuhn – Tucker conditions, KTC), as an approximation of the Lagragian function.

The optimization of the process used in Colombia for the production of biodiesel from palm oil is a problem that can be represented as follows [24]:

Maximize: $$f(x) \quad (6)$$

Subjected to: $$g(x) \geq b \quad (7)$$

The objective function is maximizing the biodiesel production, and the restrictions to this function, are the equations of the equipment involved in the production process. In addition to the inequality constrains due to the 99.6 wt% purity required by the FAME market.

The Lagragian function for this problem is:

$$L(x,\lambda) = f(x) + \lambda^T (g(x) - b) \quad (8)$$

and the KTC

$$\nabla_x L = \nabla f(x) + \sum_{i=1}^{m} \lambda_i \nabla g_i(x) = 0 \quad (9)$$

and

$$g(x) \geq b \quad (10)$$

6 Results and Discussion

6.1 Optimization

To evaluate the influence of certain parameters with a high degree of uncertainty in a design to be optimized, Himmelblau et. al. (2001) recommends the use of a sensitivity analysis.

The sensitivity analysis in non-linear programming (NLP) is a numerical or graphical indication of how a variation of the problem variables has a repercussion on the objective function [24]. The variation for this particular case of study was chosen as 80% to 120% of the value obtained by the rigorous calculation.

6.1.1 Pre optimal analysis

The objective of the pre optimal analysis is to determine the most influential parameters in the objective function. In other words this type of analysis allows the researcher to identify those aspects that have an important influence in the biodiesel production.

In the optimization's first stage, variations on process methanol flow, FAME bottoms flow rate (T-401), and the extraction solvent flow rate were performed to find the optimum operational conditions. As shown by Figure 9, the bottoms flow rate is a very sensitive variable showing a marked linear tendency with variation over the variable to be optimized.

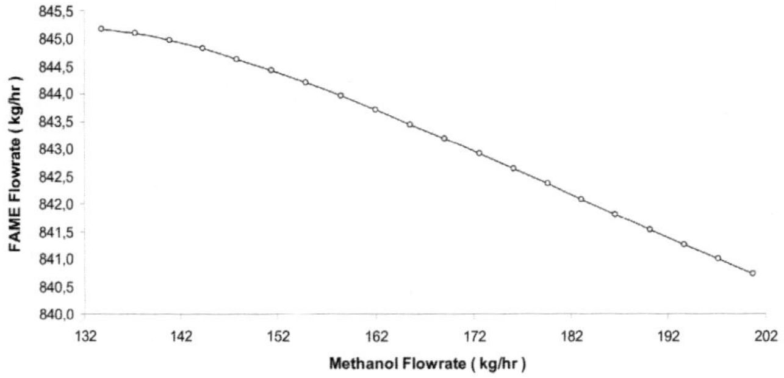

Figure 7 Sensitivity Analysis Methanol Flow rate vs. FAME Flow rate.

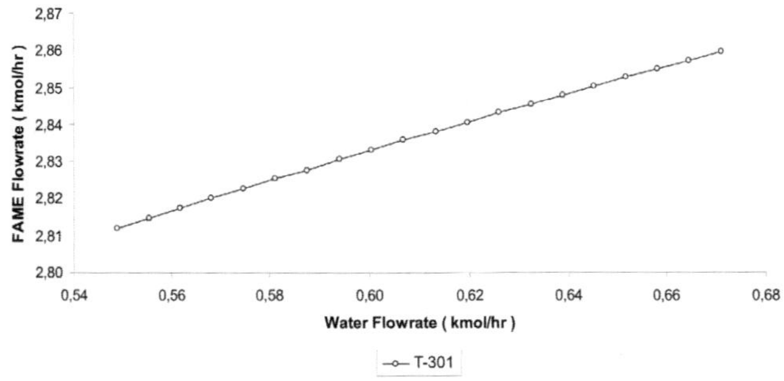

Figure 8 Sensitivity Analysis Tower T-401 Bottoms Flow rate vs.

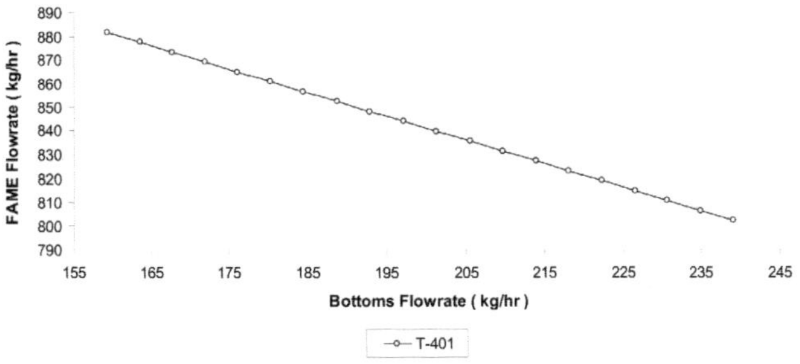

FAME Flow rate.

Figure 9 Sensitivity Analysis Water Flow vs. FAME Flow rate.

Even though variations on the quantity of methanol affect significantly biodiesel production, the changes conducted on the extraction tower T-301 water stream, as the extraction solvent flow rate, reported important results.

6.1.2 *Post optimal analysis*

Once the optimization by means of the variation of methanol flow, the solvent extraction column and the fluctuations on the bottoms flow in the FAME purification tower were carried out; a second sensitivity analysis was implemented to determine if the values obtained by the software were local optima or global optima. In other words, if the value found is a maximum value or there are other alternatives for better results; this idea can be illustrated by Figure 10, Figure 11 and Figure 12.

$$Increase\% = \frac{Optimum - Initial}{Initial} \times 100 \qquad (7)$$

The final value for the optimization of the biodiesel production plant analyzed in this paper was 1,047.54 kg/hr. This means that one kilogram of palm oil produces approximately a kg of biodiesel.

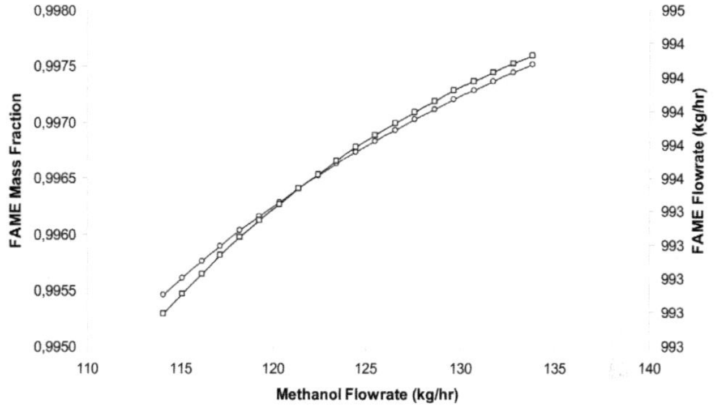

Figure 10 Sensitivity Analysis Methanol Flow vs. FAME Flow rate.

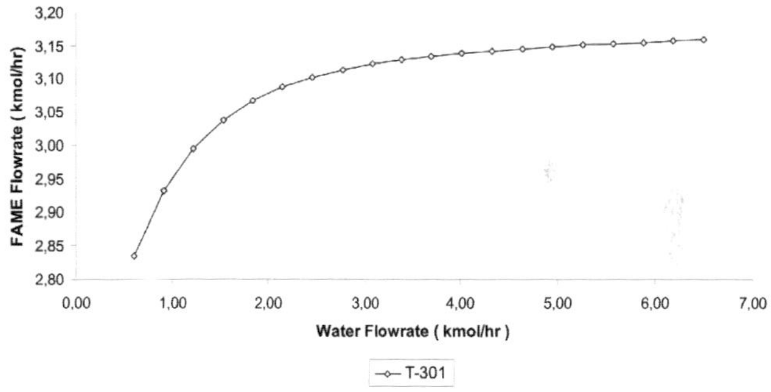

Figure 11 Sensitivity Analysis Water Flow vs. FAME Flow rate.

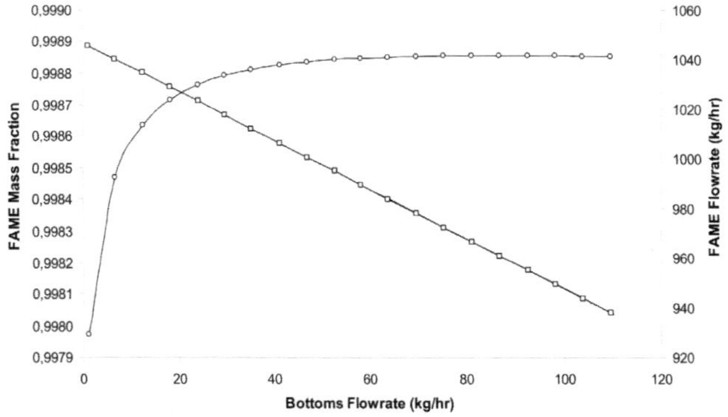

12 Sensitivity Analysis Tower T-401 Bottoms Flow rate vs. FAME Flow rate.

Comparatively, this article presents an increase in FAME production over the rigorous calculation of 24%, while it enhanced the short calculation by 10% and Zhang et. al. (2003) publication by 5%.

For this case of study a very specific analysis was effectuated for methanol and bottoms flow rate. Figure 10 illustrate that the optimum value for this variables has been achieved since changes on the methanol flow rate did not represent important variations on ether the FAME flow or FAME mass fraction. On the other hand, Figure 12 show that for variations on the bottoms flow there were important fluctuations on FAME flow rate without disturbing importantly the purity of the product.

6.2 Costs

A design has to represent a profitable plant. Initially, enough capital has to be available for construction and its adequacy, as a key aspect of the industrial unit.

Estimates on equipment cost and other related capital investment costs, play a crucial roll on the selection of design alternatives [25].

Costs Analysis for the Biodiesel Production Process

Main Process Units	Parameters	
Transesterification	Volume, m3	4,86
(Reactor CSTR)	Material	Carbon Steel
	Cost, US	$ 1.200.000
Recovery	Diameter, m	0,32
Methanol	Material	Carbon Steel
(Distillation Column)	Height, m	6
	Cost, US	$ 64.800
Washing column	Height, m	4,8
(Extraction Column L-L)	Diameter, m	1,1
	Adjusting Factor (Pressure)	1,6
	Adjusting Factor (Material)	1,0
	Cost, US	$ 24.000
Purification	Diameter, m	0,9
FAME	Material	Carbon Steel
(Distillation Column)	Height, m	4,8
	Cost, US	$ 19.200
Glycerol	Diameter, m	0,47
(Distillation Column)	Material	Carbon Steel
	Height, m	4,8
	Cost, US	$ 12.000
Heat Exchangers		$ 4.000
Pumps		$ 45.000
Other Costs (Separator, Vacuum System)		$ 46.000
Total Basic Module Cost, C_{MBO}		$ 1.415.000
Total Bare Module Cost, C_{MB}		$ 1.768.750
Contingency Fee, $C_C = 0.18C_{MB}$		$ 318.375
Total Module Cost, $C_{TM} = C_{MB} + C_C$		$ 2.087.125
Auxiliary Facility Cost, $C_A = 0.3C_{MB}$		$ 530.625
Fixed Capital Cost, $C_{FC} = C_{TM} + C_A$		$ 2.617.750
Working Capital, $C_T = 0.15C_{FC}$		$ 392.663
Total Capital Investment, $C_{TC} = C_{FC} + C_T$		$ 3.550.498

Table 1 Costs Analysis for the Biodiesel Production Process.

The costs in Table 1 are listed in United States dollars. Costs for the equipment were updated through the Marshall & Swift cost index obtained in the Chemical Engineer magazine, where I_{2002} = 1,104.2 y I_{2006} = 1,302.3 [27].

Turton et. al. (1998) mentioned that this type of estimate was a + 30% and -20% precision. Thus, the results don't reflect a strict capital cost analysis; instead they are a very useful comparative parameter.

7 Results Analysis

Two important characteristics were identified while comparisons were made between the results obtained for the short calculation and the base case simulation: the reproductive and repetitive character of the exercise. The reproductive character is associated with the fact that the simulation could be replicated and the repetitive factor is supported on the 82% concordance found among the Zhang et. al. (2003) article's results and this case of study outcome. This discrepancy could be related with the differences in the thermodynamical parameters stored within each one of the software used for the particular papers.

A detailed analysis on Table 2 reveals the potential for improvement that has the technical evaluation developed by Zhang. et. al. (2003) confronted with this case of study. Consequently, a 3.8% increase on the reactor conversion has been related mainly with a 20 ºC modification effectuated within the base case, on the reactors temperature. Moreover, the methyl ester fraction increased in 5 wt% as a response on the 6 times increment experienced by the solvent on the liquid – liquid extraction tower.

A closer approach of this important table exposes two of the most important concepts related to optimization: local and global optimum. A local optimum has an objective value that is better than the one of any nearby feasible solution, while global optimum is a feasible solution that has the best objective value. Since fluctuations on the variables analyzed did not implicate considerable changes, on the optimum biodiesel production found this value can be categorized as a global optimum.

Comparative Analysis for the Biodiesel Production Process

Parameters	Zhang. et al.	Velosa. et. al. P I	Velosa. et. al. P II
Reaction Temperature, C	60	80	80
Reactor Yield	95,0%	99,8%	99,8%
Methanol Purity, wt%	1,000	1,000	1,000
Water Flow, kg/h	11	66	66
FAME Extraction, wt%	0,946	0,994	0,993
Glycerin Purity, wt%	0,850	0,994	0,995
FAME Purity, wt%	0,997	0,998	0,998
FAME Production, kg/h	999,88	1048,06	1047,54
Cost, US	$ 3.144.010	$ 3.550.498	$ 3.550.498

Table 2 Comparative Analysis for Different Biodiesel Production Processes.

The reactor as the plant's hearth and one of the most important units of this complicated process should have a detailed analysis. The operative pressure and temperature range has to be established by means of a kinetic analysis as well as the physical characteristics of the reactants. The temperature range was determined through the graphical analysis of Figure 16, which illustrates the Ln (k) variation against 1/T. As the production of diglyceride increases and the temperature augments the line that represents the diglyceride to monoglyceride reaction is located above the line which represents the diglyceride to triglyceride reaction, obtaining enough monoglyceride in order to produce methyl oleate (methyl ester). Furthermore, the pressure range was determined as to avoid methanol evaporation at the established temperature. At 400 kPa methanol's boiling point is 104.7 °C.

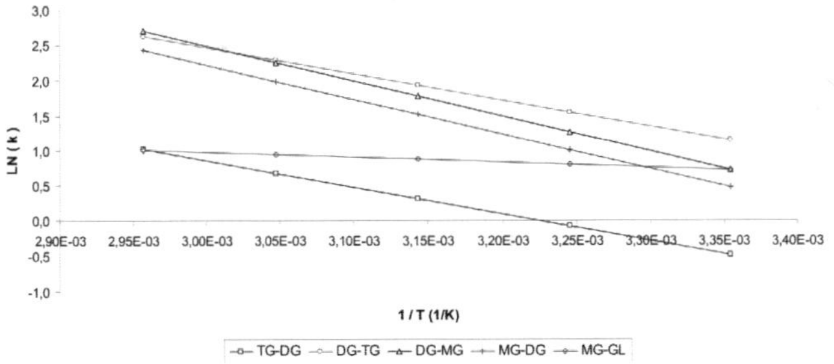

Figure 13 Kinetic Analysis of the Transesterification Reaction.

The optimization of biodiesel production and the glycerin purity, finding the operation and design conditions that made the process a competitive one is the most important contribution of this article. In this case of study glycerin surmounts the expectations since it presented a 17% increase compared with the base case. However, Zhang et. al. (2003) has reported that up to a 92 wt% purity is possible by means of a distillation column; this type of variation can be related to the thermodinamical model, as well as the prevailing discrepancies among the simulation software. Moreover, biodiesel production increased significantly 5% compared with the base case.

At the optimum point, variables as the water flow and the distillate flow in FAME's purification column have an important impact over biodiesel production, in other words fluctuations around this variables implicate significant changes on the objective function.

An important variable affecting the conversion of ester is the alcohol to triglyceride molar ratio; higher molar ratios result in greater ester conversion in a shorter time. For this particular work the optimum molar ratio was established at 7:1.

One of the most important process operations is the liquid – liquid extraction column, because it allows the separation of methanol, glycerol (as a secondary product) and the main product. At this unit, the water flow has a significant influence on the process which is directly related with the column exit methyl ester fraction. A close insight on this last paragraph shows that it is necessary to use 66 kg/h of water in order to obtain a 99.3 wt% purity of FAME, presenting an asymptotic tendency at a value of 3.13, in other words an increase on the amount of water added would not represent an important increase on the purity of the main product.

A detailed analysis on the distillation columns permits an evaluation of the most important variables established by Zhang et. al. (2003). On this type of industries profitability could be allocated to on the ability to encounter new sources of revenue; this is the case of glycerol and mainly its purity.

FAME and glycerol are susceptible to the thermal decomposition above 250 and 150 °C, respectively [10]. Glycerin purity could not be above a value of 95.3 wt% with a unique separation unit at a pressure of 40 kPa, since the temperature obtained at the bottom of the tower would be 149 °C, value that represents an imminent denaturalization risk. Nonetheless, if the pressure is decreased to 5 kPa the main product purity reaches a value of 99.4 wt%, with a bottoms temperature of 143.4 °C.

The biodiesel purification tower defines the products′ quality through the distillate flow. As the relation of those two variables is direct, an increment on the distillate flow would permit an increase on the biodiesel purity until the quality and design constrains are reached; this procedure would structure the global optimum finding. In this particular case, adding the distillate flow as a variable to be optimized represented a 13% increase on FAME production comparatively.

The economical evaluation has been a great complement to the simulation and optimization processes, since it reduces the limitations between the academic exercise and reality and permit's a starting point for the election of the variables to be optimized from a financial and not only a technical, point of view.

Zhang et. al. (2003) determined the cost of a biodiesel production plant with the same characteristics described by this article, obtaining an estimated investment of U$ 3,144,010. On that article the biodiesel production was established at 8,000 tons/year, the reactor conversion was set at 95% and the glycerin purification costs were not included. However, the total cost of this research production mill was U$ 3,550,498, with a 99.8% reactor conversion, as well as a glycerin purity of 99.4 wt%. In terms of production, this 5% increment will represent approximately a 330.322 U$/year profit (biodiesel prices obtained from Zhang et. al. (2003) and Castañeda et. al (2006)).

The cost difference for the two plants is based principally on the reactor volume increase, in order to have a 99.8% conversion, in addition to the material augmentation that this change represents. It is important to mention that this conversion value is very similar to the one presented by companies as Lurgi [30].

As the technical aspect evolves in order to find the optimal production conditions before the start up of the production plants, the National Government and the private industry develop projects that facilitate the conditions necessary to meet the politics of the laws.

The National Government projections for the following years include plantations of one million African palm hectares, focused on the biodiesel production for national consumption as well as its exportation, producing approximately 150,000 barrels/day; this scheme will generate an estimate of one million direct jobs, and an important alternative for the illicit crops substitution.

The legislature has joint efforts as a strategic ally providing an important set of advantages. These advantages go from tax reduction (Law 939/04), to duty free territories (Decreto 383/07), as to promote investors to support this type of industry. In addition, based on the region characteristics the Interamerican Development Bank (IDB) will finance biofuels production projects in Latin America.

Biofuels are one of the most important economical and energetic alternatives that Colombia has experimented since the Cusiana and Cupiagua discovery. It is time to use the condition of an agricultural country, in order to establish independence on non renewable combustion sources, playing a main roll on the technological development and implement actions that allow an increase in FAME production efficiency.

8 Conclusions

The continuous biodiesel production process using palm oil, as raw material, by means of homogenous catalysis was simulated obtaining satisfactory 8,400 tons/year production. The bibliographic review allowed the definition of the thermodynamical models and kinetic models necessary to accomplish conditions as real as possible. The optimization part of this study incremented the production in 5% compared with the base case published by Zhang et. al (2003).

To develop an academic analysis, the simulation tool chosen for the production mill modeling was Aspen Plus®, demonstrating among all its benefits the replicability and reproducibility character, establishing the bases in order to be able to compare the results obtained with the ones described by the base case.

The use of three different thermodinamical activity models as: NRTL, UNIQUAC and UNIFAC-Dortmund have been shown to reduce the limitations between the academic exercise and reality, since it reflects the experimental data and the complete immiscibility of the main compounds.

The relationship between the technical concepts and the economical factors is decisive when a project is evaluated, since those two perspectives are directly involved on its viability.

A projects economical evaluation is a great complement to a technical approach, since reduces the limitations between the academic exercise and reality and allows a starting point for the election of the variables to be optimized.

The variables that have a mayor effect on this case of study are the water flow and the distillate flow on the FAME purification unit. Consequently, the variation and inclusion of this parameters on the software incremented the base case production by 5% and 24% compared to the rigorous calculation; these values represent a utility increase of 330.322 U$/year, taking into account that the methyl ester prices were calculated as to meet the equilibrium point (zero utility).

The reactor conditions were fixed at 80 °C and 400 kPa, since those are the most adequate for the plants correct performance.

Interestingly, the glycerin purity can not be set over 99.4 wt% purity without having a thermical denaturalization possibility if the columns pressure is 5 kPa.

The biodiesel production cost for this study is U$ 3.550.498, with a 99.8% reactor conversion, a 99.8 wt% FAME purity and a 99.4 wt% glycerin purity.

Nomenclature

A *Methanol.*
T *Triglyceride.*
D *Diglyceride.*
M *Monoglyceride.*
G *Glycerol.*
E *Methyl Ester.*
R *Alcoholic radical.*
R´ *Triglyceride radical.*
wt% *Weight percentage.*
K *Equilibrium constant.*
k_1 *Reaction constant triglyceride-diglyceride.*
k_2 *Reaction constant diglyceride-monoglyceride.*
k_3 *Reaction constant monoglyceride-glycerol.*
k_4 *Reaction constant diglyceride-triglyceride.*
k_5 *Reaction constant monoglyceride-diglyceride.*
k_6 *Reaction constant glycerol-monoglyceride.*
λ *Lagrange multipliers.*

Acknowledgements

The author acknowledges the collaboration of companies as Cenipalma, Fedepalma and Cemex. The Los

Andes University professors Dr. Jorge Mario Gomez, Ing. Ruben Fajardo and Ing. Ivan Gil. Furthermore, his family, Natalia Olarte and friends.

9 References

[1] Knothe. G.; Krahl. J; Gerper. J. The Biodiesel Handbook., 2004. AOCS Press, Champaign, Illinois. (Chapters 1 and 2).

[2] Ma. F.; Hanna. M. Biodiesel production a review. *Bioresourse Technology 70*. 1999.

[3] Lee, D. H.; Lee, D. J. Biofuel economy and hydrogen competition. *Energy & Fuels 22*. 2008.

[4] Cunha, R. Potencial for producing bio-fuel in the Amazon deforested areas. *Biomass and Bioenergy 26*. 2004.

[5] Piedraita. C. Boletín Estadístico de Minas y Energía. Ministerio de Minas y Energía. Unidad de Plantación Minero Energético 2006. *Recovered September 25th form the web page: http://www.upme.gov.co/Index2.htm*.

[6] Rincón. S.; Cuellar. M.; Cadena. A. Estudio Prospectivo para el Desarrollo de la Oleoquímica del Aceite de Palma en Colombia (Cenipalma 2002).

[7] Federación Nacional de Cultivadores de Palma de Aceite (Fedepalma). Balance Económico. Unidad de Economía y Comercio de Fedepalma. March 2007.

[8] Zhang, Y., 2003. Design and economic assessment of Biodiesel production from waste cooking oil. Part 2. M.A.Sc. thesis, Department of Chemical Engineering, University of Ottawa.

[9] Zhang, Y., 2003. Design and economic assessment of Biodiesel production from waste cooking oil. Part 1. M.A.Sc. thesis, Department of Chemical Engineering, University of Ottawa.

[10] Velosa, F.; Gomez, J. M. Simulation and Optimization of the process used in Colombia for the production of Biodiesel from palm oil. July 2007.

[11] You, Y. D. Economic cost analysis of biodiesel production: case in soybean oil. *Energy and Fuels 22*. 2008.

[12] West, A. H. Assessment of four biodiesel production process using HYSYS.Plant. *Bioresource Technology, 2008.*

[13] Freedman, B.; Pryde, E. H.; Mounts, T. L. Variables affecting the yields of fatty esters from transesterified vegetable oils. *J. Am. Oil Chem. Soc.* 1984, *61* (10), 1638.

[14] Vicente, G.; Martinez, M.; Aracil, J.; Esteban, A.; Kinetics of sunflower oil methanolysis. *Ind. Eng. Chem. Res.* 2005, 44, pages 5447-5454.

[15] Noureddini. H.; Zhu.D. Kinetics of transesterification of Soybean Oil. JAOCS, V. 74, No 11, 1997, pages 1457 – 1463.

[16] Darnoko, D.; Cheryan, M. Kinetics of palm oil transesterification in a batch reactor. *J. Am. Oil Chem. Soc.* 2000, *77* (12), 1263.

[17] Jordan. V.; Gutsche. B. Development of an environmentally benign process for the production of fatty acid methyl esters. *Chemosphere 43*, 2001, pages 99 – 105.

[18] Germen. J. A Workshop for the 2005 Biodiesel Conference and Expo. *Biodiesel Production Technology*. February, 2005.

[19] Seider, W.; Seader, J.D.; Lewin, D. Product and process design principles. Wiley. 2004. Second Edition.

[20] Antolin. G.; Tinaut, F.; Briceno. Y.; Castano. V.: Perez, C.; Ramirez, A. Optimization of Biodiesel production by sunflower oil transesterification. *Bioresourse Technology* 83. 2002.

[21] Negi. S.; Sobotka. F.; Kimmel. T.; Wozny. G.; Schomacker. R. Liquid – Liquid Phase Equilibrium in Glycerol – Methanol – Methyl Oleate and Glycerol – Monoolein – Methyl Oleate Ternary Systems. *Ind. Eng. Chem. Res, 2006, 45*. Pages 3693 – 3696.

[22] Carlson. E. Don't Gamble With Physical Properties For Simulations. Aspen Technology. Inc. 1996. *Chemical Engineering Progress.*

[23] Reid. R.; Prausnitz. M.; Poling. E. The Properties of Gases & Liquids. McGraw Hill, 1988. Fourth Edition.

[24] Himmelblau. D.; Edgar. T. Optimization of Chemical Processes. McGraw Hill, 2001. Second Edition.

[25] Peters. M.; Timmerhaus. K.; West. R. Plant Design and Economics for Chemical Engineers. McGraw Hill, 2001. Fifth Edition.

[26] Turton, R., Bailie, R.C., Whiting, W.B., Shaeiwitz, J.A., 1998. Analysis, Synthesis, and Design of Chemical Processes. Prentice Hall PTR, New Jersey.

[27] Chemical Engineering. Industrial Electrochemistry: A Look at the Plus Side. May 2007. Volume 114, Issue 5.

[28] R. Perry.; D. Green.; Maloney. Manual del Ingeniero Químico, 2001. Volume 1. McGraw Hill. Seventh Edition.

[29] Castañeda. Z.; Narváez. P.; Rincón. S.; Sánchez. F. Simulación de un Proceso para la Producción de Esteres Metílicos de Aceite de Palma.

[30] Confidential Document. Cenipalma. Reviewed June 2007.

394c Process Development to Reduce Minor Component Content In Biodiesel

Haiying Tang, Rhet Joseph De Guzman, Steven O. Salley, and K. Y. Simon Ng. Department of Chemical Engineering and Materials Science, Wayne State University, 5050 Anthony Wayne Drive, Detroit, MI 48202

Biodiesel is a renewable and environmentally friendly alternative fuel, which is comprised of mono-alkyl esters of long chain fatty acids derived from vegetable oils or animal fats. However, filter plugging and engine failure have taken place in vehicles using biodiesel blends due to precipitate formation at low-temperature. The majority of these precipitates can be attributed to either steryl glucosides or monoglycerides found in biodiesel. These dispersed particles increase the filtration time, and also promote the crystallization of other compounds. Up until now, no ASTM standard analytical methods to determine steryl glucosides in biodiesel exist. The objective of this work is to evaluate and develop processing strategies to reduce steryl glucosides, and monoglycerides content in biodiesel, as well as to develop a robust analytical method to determine their content in biodiesel.

A number of processing techniques were evaluated: 1. ambient filtering; 2. cold filtering; 3. water degumming; 4. absorbent treatment; and 5. vacuum distillation. In each processing technique, the amount of sterol glucosides and glycerin in biodiesel before and after the processing treatment was determined by gas chromatography - flame ionization detector (GC-FID). Moreover, cold flow proprieties, viscosity, acid number, and oxidative stability, flash point, and Cetane number of biodiesel were evaluated.

Developing a cost-effective processing technique to reduce sterol glucosides and glycerin can result in reduced precipitate formation, and minimize filter plugging problems for the automotive fuel delivery system. By improving its quality, biodiesel will become more competitive on the market, and its acceptance to the public will be enhanced.

394d Study of Antioxidants on the Oxidative Stability of Biodiesel

Haiying Tang, **Rhet Joseph De Guzman**, Steven O. Salley, and K. Y. Simon Ng. Department of Chemical Engineering and Materials Science, Wayne State University, 5050 Anthony Wayne Drive, Detroit, MI 48202

A significant problem associated with the commercial acceptance of biodiesel is poor oxidative stability. Previous studies indicate that different types of biodiesel have different natural levels of oxidative stability. Moreover, the effect of both natural and synthetic antioxidants on biodiesel stability differs depending on composition. Therefore, understanding the mechanism of antioxidant activity and developing new antioxidants to improve fuel oxidative stability are important.

This study develops a robust analytical method to determine natural antioxidant content in biodiesel, and investigates the effect of composition and minor components, such as sterols, natural antioxidant, and free and bound glycerin, on the effectiveness of synthetic antioxidant. Additionally, the effect of selected antioxidants on oxidative stability of biodiesel under long-term indoor and outdoor storage will be described.

Catalytic Deoxygenation of Oleic Acid over Ceria-Zirconia Catalysts

Boonyarach Kitiyanan[a], *Charothon Ung-jinda*[a], *Vissanu Meeyoo*[b,1], *Thirasak Rirksomboon*[a], *and Pramoch Rangsunvigit*[a]
[a] *The Petroleum and Petrochemical College, Chulalongkorn University, Bangkok, Thailand*
[b] *Centre for Advanced Materials and Environmental Research, Mahanakorn University of Technology, Bangkok, Thailand*

Abstract

The catalytic deoxygenation of oleic acid was studied over ceria-zirconia mixed metal oxide (CZMMO), $Ce_{1-x}Zr_xO_2$ (x = 0, 0.25, 0.50, 075 and 1.0), and Fe, Ni, Cu and Rh loaded on the CZMMO catalysts. Experiments were carried out in a continuous flow reactor within the temperature range of 325-425°C. The gaseous products mainly are CO and CO_2 while the main compositions in liquid products are C_{17} and C_{18} hydrocarbons. The difference in the product distribution was insignificant for all types of the catalysts used in this work. Surprisingly, $Ce_{0.75}Zr_{0.25}O_2$ exhibits the highest catalytic activity for the deoxygenation of oleic acid with high selectivity toward C_{17} hydrocarbons. The catalytic activities of metal supported on $Ce_{0.75}Zr_{0.25}O_2$ catalysts are lower than that of $Ce_{0.75}Zr_{0.25}O_2$. This is due to a product hindrance. A kinetic study of deoxygenation of oleic acid over $Ce_{0.75}Zr_{0.25}O_2$ support was further carried out. First order kinetic was well fitted for the deoxygenation of oleic acid and the estimated activation is ca. 23 kcal mol^{-1}.

Introduction

It has been reported that the liquid products from pyrolysis of sewage sludge consist mainly of a long chain carboxylics [1]. This fraction could be upgraded by deoxygenation reaction and potentially be used as a diesel fuel additive for the cetane number booster [2]. Therefore, the catalytic deoxygenation unit seems reasonable to be implemented in the sequence with the pyrolysis unit. The deoxygenation of carboxylic acids such as acetic acid [4], benzoic acid [5, 6] and octanoic acid [7] has been studied over the several types of the metal oxide catalysts. From these reports, the metal oxide catalysts illustrate good catalytic activity. Therefore, our aim is to study the deoxygenation of oleic acid as a model of high carbon number carboxylic acid obtained from sewage sludge. The ceria-zirconia mixed metal oxide (CZMMO) catalysts with various compositions were tested for their catalytic activity here under steady-state conditions in a continuous flow reactor within the temperature range of 325-425°C. Moreover, the kinetic studies of deoxygenation were also carried out.

---

[1] Corresponding author Tel.: +66(0)-2988-4043; fax: +66(0)-2988-4039.
Email address: vissanu@mut.ac.th (V. Meeyoo)

Results and discussion

Deoxygenation of oleic acid (OA), $C_{17}H_{35}COOH$, can take place via two pathways. The first one is the direct decarboxylation forming C_{17} hydrocarbons and the second one is the ketonization forming symmetrical ketones of C_{35} and CO was then removed by the decarbonylation converting to aliphatic hydrocarbons of C_{17} and C_{18}. The gaseous products from thermal and catalytic deoxygenation of OA are CO and CO_2 and the liquid products contain hydrocarbons ranging from C_6-C_{18}. The conversion of OA was calculated from CO and CO_2 production. The effect of reaction temperature was illustrated in Fig. 1. For example at 385°C, the highest OA conversion of 20.8% was observed for $Ce_{0.75}Zr_{0.25}O_2$ followed by 15.7% for CeO_2. For $Ce_{0.50}Zr_{0.50}O_2$, $Ce_{0.25}Zr_{0.75}O_2$ and ZrO_2, the OA conversions were comparable at 10.9 and 12.8 and 11.9%, respectively. The results were found to relate with the redox property of the CZMMO catalysts. This is confirmed by the CO-TPR results, where $Ce_{0.75}Zr_{0.25}O_2$ showed the highest reducibility [8]. The liquid products are hydrocarbon ranging from C_6 to C_{18} and the main products are C_{17} and C_{18} hydrocarbons. The composition distribution in liquid fraction is similar in all catalysts studied. However, $Ce_{0.75}Zr_{0.25}O_2$ and CeO_2 was found to be more selective toward C_{17} and C_{18} hydrocarbons.

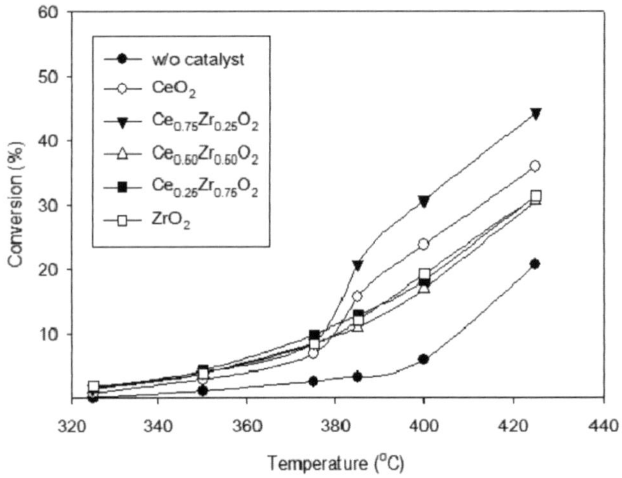

Fig. 1. Conversion versus temperature of deoxygenation reaction of oleic acid

Effects of doped metal were studied on the $Ce_{0.75}Zr_{0.25}O_2$. CO and CO_2 evolution rate from the deoxygenating reaction of oleic acid over Fe, Ni, Cu and Rh supported on $Ce_{0.75}Zr_{0.25}O_2$ at 385°C were shown in Fig. 2. CO production rates were found to be constant at ca. 3.0×10^{-6} mol s^{-1} g^{-1}_{CAT} for all catalysts. The maximum CO_2 production rate was ca. 2.52×10^{-5} mol s^{-1} g^{-1}_{CAT} for $Ce_{0.75}Zr_{0.25}O_2$ catalysts. Doping of metals on the $Ce_{0.75}Zr_{0.25}O_2$ was found to reduce the OA deoxygenation rate. This might be due to CO produced from the reaction may strongly adsorb on the metal surface, leading to a loss in the catalytic activity.

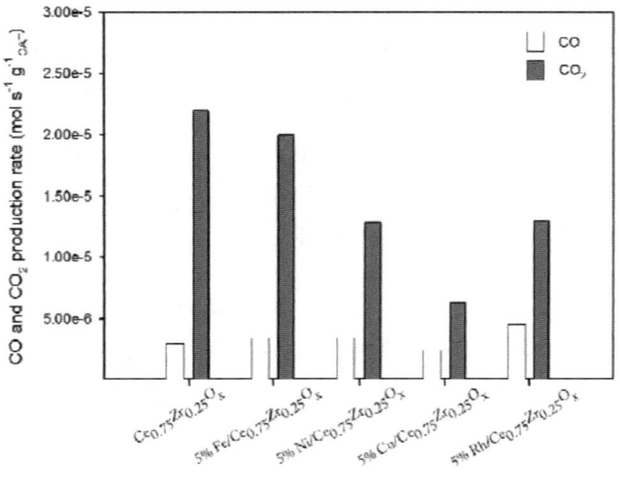

Fig. 2. CO and CO_2 evolution rate from the deoxygenation of oleic acid over metal supported on $Ce_{0.75}Zr_{0.25}O_2$ at 385°C

A first order kinetic model was applied for the deoxygenation of oleic acid over $Ce_{0.75}Zr_{0.25}O_2$. The initial rate and rate constant are summarized in Table 1. And subsequently the activation energy (E_a) were determined to be 23 kcal mol^{-1} and the Arrhenius constant (A) is 1.3×10^9 l h^{-1} g_{CAT}^{-1}.

Table 1. Initial rate of reaction and reaction constant at 375, 385 and 400°C

T (°C)	r_o (mol s^{-1} g_{CAT}^{-1})	k (l h^{-1} g_{CAT}^{-1})
375	8.29×10^{-7}	1.14
385	1.05×10^{-6}	1.45
400	1.60×10^{-6}	2.20

Conclusions

In conclusion, the deoxygenation of oleic acid can be achieved via catalytic decarboxylation. The products obtained are mainly C_{17} and C_{18} for the case of ceria-zirconia catalysts. With the presence of metals, the deoxygenation is achieved via decarboxylation and ketonization. $Ce_{0.75}Zr_{0.25}O_2$ was found to be the most active and stable catalyst for oleic acid deoxygenation. A first order kinetic model was found to fit perfectly with experimental data. The calculated activation energy is ca 23 kcal mol^{-1}.

Acknowledgements

This work was supported by the Thailand Research Fund (under Waste-to-Energy project and Royal Golden Jubilee Ph.D. Program: Grant 0061/45), and the Research Unit for Petrochemical and Environmental Catalysts, Ratchadapisek Somphot Endowment Fund, and the National Center of Excellence for Petroleum, Petrochemicals and Advanced Materials, Chulalongkorn University.

References

[1] C. Jindarom, V. Meeyoo, T. Rirksomboon, P. Rangsunvigit, (2007), Thermochemical decomposition of sewage sludge in CO_2 and N_2 atmosphere, Chemosphere, 67, 1477–1484.

[2] S. Zhang, Y. Yan, T. Li, Z. Ren, (2005), Upgrading of liquid fuel from the pyrolysis of biomass, Bioresource Technol., 96, 545–550.

[3] K. Parida, H. K. Mishra, (1999), Catalytic ketonisation of acetic acid over modified zirconia 1. Effect of alkali-metal cations as promoter, J. Mol. Catal A, 139, 73-80.

[4] M.W. de Lange, J.G. van Ommen, L. Lefferts, (2001), Deoxygenation of benzoic acid on metal oxides 1. The selective pathway to benzaldehyde, Appl. Catal. A: General, 220 41-49.

[5] M.W. de Lange, J.G. van Ommen, L. Lefferts, (2002), Deoxygenation of benzoic acid on metal oxides 2. Formation of byproducts, Appl. Catal. A: General, 231, 17-26.

[6] F. Billaud, Y. Guitard, A.K. Tran Minh, O. Zahraa, P. Lozano, D. Pioch, (2003), Kinetic studies of catalytic cracking of octanoic acid. J. Mol. Catal. A: General, 192 281-288.

[7] S. Pengpanich, V. Meeyoo, T. Rirksomboon, K. Bunyakiat, (2002), Catalytic oxidation of methane over CeO_2-ZrO_2 mixed oxide solid. Appl. Catal. A: General, 234 221-233.

[8] M. Thammachart, V. Meeyoo, T. Rirksomboon, S. Osuwan, (2001), Catalytic activity of CeO_2-ZrO_2 mixed oxide catalysts prepared via sol-gel technique: CO oxidation. Catal. Today, 68 53-61.

394f Product and Process Effects of Alternative Solid Fuels Usage at a Portland Cement Manufacturing Site

Steve R. Duke[1], Srikanth Akkapeddi[2], Dustin Swart[2], Anton Schindler[2], and Don Stafford[3]. (1) Chemical Engineering, Auburn University, 212 Ross Hall, Auburn University, AL 36849-5127, (2) Civil Engineering, Auburn University, 221 Harbert Engineering Center, Auburn University, AL 36849-5127, (3) US East Business Unit, Lafarge North America, 12735 Morris Road, Extension, Suite 300, Alpharetta, GA 30004

Portland cement manufacturing involves the combustion of solid fuels along with raw materials to produce clinker. Temperatures of 1,500 °C are needed to carry out reactions for cement chemistry. Fuel costs and environmental impact concerns encourage the cement industry to explore alternatives to the use of fossil fuels. Potential alternative fuels must provide adequate and economical energy but they must also result in products of combustion that have negligible impact on cement quality.

Results will be presented from a multiyear study (supported by the Department of Energy) in which portland cement was produced at a full-scale cement plant with several 3-day trial burns of various alternative fuels along with coal. The fuels combinations investigated were: 1) coal only, 2) coal and tires, 3) coal, tires, and waste plastics, 4) coal, tires, and broiler litter, 5) coal, tires and switch grass, and 6) coal, tires, and wood chips. Samples of raw materials, fuels, cement kiln dust, clinker, cement, and emissions were collected systematically during each trial burn. Chemical compositions, physical characteristics, and mechanical properties were obtained for the samples. Chemical analyses showed that the primary chemical compounds of the cements exhibited no significant changes during the trial burn periods. Cement and concrete properties for each trial burn showed no significant changes in drying shrinkage development, splitting tensile strength, and permeability. The compressive strength of concrete from waste plastics and broiler litter burns showed an increase relative to the other burns, though it is not possible to attribute this result exclusively to the use of these fuels. The NOx emissions were the lowest for the coal only and broiler litter trial burns. The VOC emissions were the lowest for the coal only and waste plastics burns. The CO emissions of the cement plant were lower for all burns that included alternative fuels.

The trial burns also revealed a number of issues and technologies which are now being addressed to enable the large scale use of alternative solid fuels for cement manufacturing. Among these are alternative fuels handling facilities and best practices and process operation and control limitations that cap the substitution rates.

The burn trials showed that all of the alternative fuels had high potential as a coal substitute because they exhibited minimal impact on product quality and in some cases had positive impact on emissions.

Biodiesel Processor Automation

Danah Hashem, Simon Huang, Joseph Pearson, and Ihab H. Farag
Chemical Engineering Dept, W315 Kingsbury Hall, University of New Hampshire,
Durham, NH 03824, 603-862-2313, ihab.farag@unh.edu

Abstract
An automated benchtop biodiesel processor was developed to simulate the operation of a processor and to implement a data acquisition and process automation approach.

1- Introduction

This project is to study the design and implementation of an automated biodiesel processor. A non-automated processor requires many man-hours for operation. An automation system can assist in adding further safety precautions while also reducing man hour time for processing. Automating the processor involves refining the processor design to make the operation more suitable for automation. The actual automation system has been studied using LabView software and a USB-NIDAQ (National Instruments Data Acquisition) interface. This arrangement is most suitable for a research environment to be used while changes in the process are being made. Based on this control system, a Programmable Logic Control (PLC) system can be developed for use with MBP's processors for customers. The PLC based systems are unique systems designed for continuous use in an industrial environment.

2- Project Goals/Objectives

The goals of the project were to build a benchtop automated system that will simulate an existing processor design with modifications necessary to make it run by computer control. And to have the processor adaptable to test alternative and novel processing techniques in the lab on the benchtop scale. This project work includes design and assembly of an automated Biodiesel processing system. The automation is done using of a PC integrated data acquisition hardware and Labview software to monitor and control the operation.

3- Methodology/Approach

The project started with a background research on manual and automated processors. Then an analysis was made of a manually operated processor design. Consideration was made for the aspects of the process that were easily adapted to automation equipment, such as replacing manual valves with solenoid actuated valves. Other parts of the design required modification of the existing system to be adaptable for automation hardware, such as the means of catalyst mixing. A flowsheet of the processor was generated detailing system design changes to accommodate automation technology. A benchtop prototype was developed. This required a chemical compatibility study. The introduction and implementation of computer control required a review of automation hardware and software. The prototype is useful in simulating a full scale automated system and in testing new processing technologies.

4- Background Research

It was important to learn the current industry practices of biodiesel production. Several organizations were contacted to find out their general processing methods including; feedstock requirements, reactor size, catalyst used, and whether the system was automated. Naturally some companies are reluctant to share information that they consider proprietary technology. The figures below show the results of our survey. The graph of percent of automated processors suggests that the industry is still maturing and there is a need for improvement in automation technology.

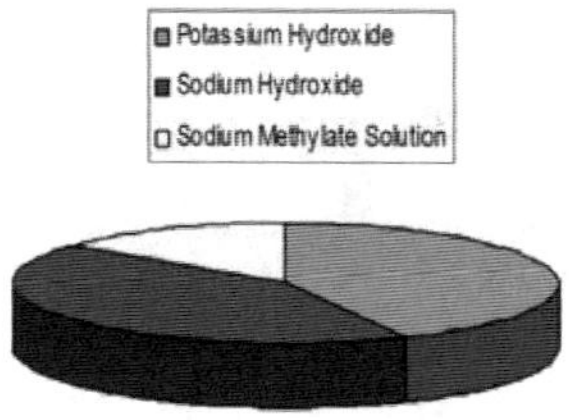

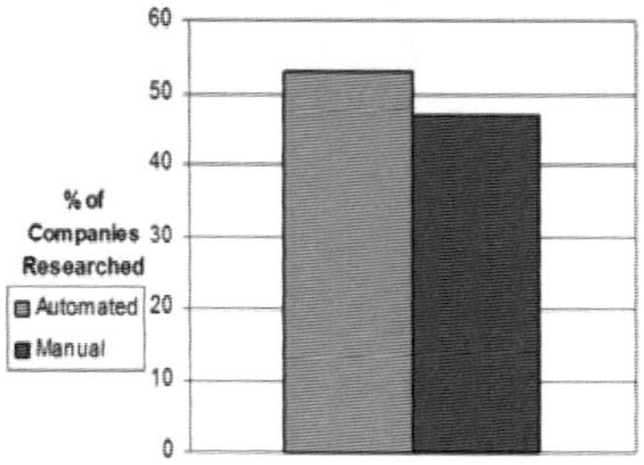

5- Chemical Compatibility

System design began with chemical compatibility research. Compatibility issues are severe in biodiesel processing. The oil and alcohol reactants are both very strong solvents, particularly the alcohol. The solvent nature of the reactants greatly limited the choices of acceptable polymers. In addition to the reactants, the finished product biodiesel also has the tendency to degrade many polymers. Furthermore, the strong alkaline catalyst is very reactive with many material options.

6- Preliminary System Design

The flowsheet of the design of the prototype processor system flow sheet is shown in figure 1.

The final system apparatus is shown in the picture in figure 2.

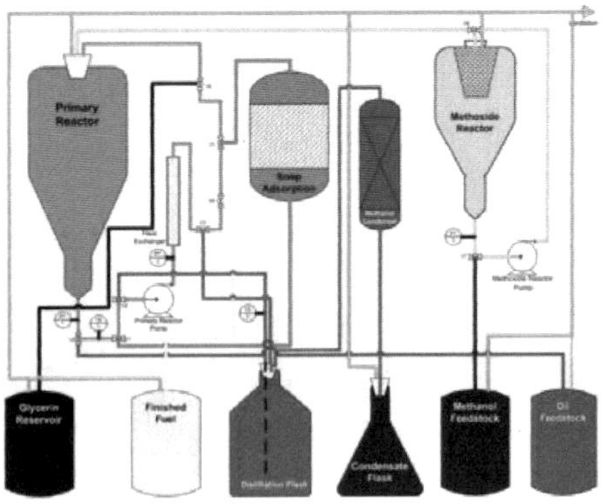

Fig 1; Prototype Automated Biodiesel Processor Flow Sheet

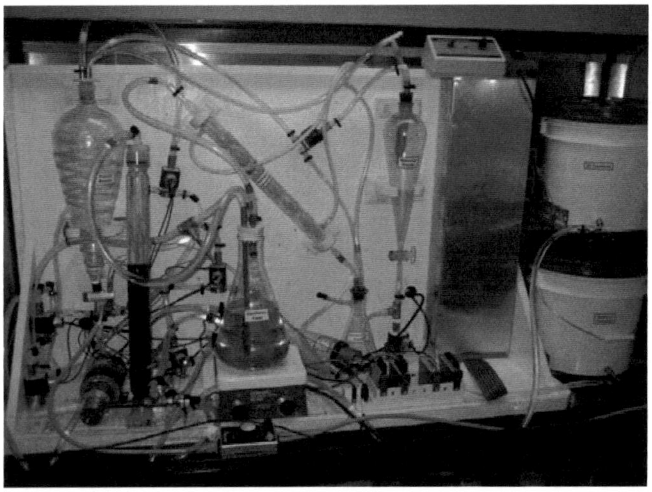

Fig 2; Prototype Automated Biodiesel Processor System

The steps in the process:

1. Oil Feed from the oil feedstock to the primary reactor by primary reactor pump.
2. Potassium Hydroxide manual addition
3. Methanol Addition from the methanol feedstock to the methoxide reactor by the methoxide reactor pump.
4. Methanol Circulation: Methanol and potassium hydroxide react and circulated.
5. Methoxide Transfer to the primary reactor to be mixed with the oil.
6. Oil and Methoxide Circulation by the primary pump. Glycerin and biodiesel are formed.
7. Drain Glycerin: Since glycerin is denser than biodiesel, it settles at the bottom of the primary reactor. Valve is opened and gravity drained glycerin into the glycerin reservoir.
8. Biodiesel Transfer to the distillation flask by the primary reactor pump.
9. Distillation and Condensation: Biodiesel is circulated with primary reactor pump and distillated to remove methanol.
10. Soap Adsorption &fatty acid removal in the soap adsorption unit
11. Finished Biodiesel Product is stored into the finished fuel tank.

The system design features two reaction vessels, a methoxide reactor and a primary reactor. Each is constructed of a Pyrex separatory funnel. The feedstock and finished product reservoirs are pails with re-sealable lids and fitted with drain ports on the bottom. The distillation flask is

Pyrex. A glass condenser column and small flask are used in the methanol recovery procedure. Two laboratory positive displacement pumps with magnetic drives are used. All of the flow streams are controlled by three way stainless steel solenoid valves. Clear tubing was used so that the process operation could be observed.

The valves are all controlled by mechanical relays and the pumps by solid state relays. Thermocouples are inline to read the temperature of the primary flow loop as well as the distillation loop. Pressure transducers are placed at the bottom of the primary reactor and the methoxide reactor vessels. These are calibrated to display the liquid level height in the vessels. The computer is interfaced with a National Instruments (NI) CompactDAQ 9172. The DAQ is outfitted with a 32 channel digital sourcing module NI9476, a 4 channel analogue input module NI9215 and a 4 channel thermocouple module with built in signal conditioning NI9211. The process control algorithm is written in NI Labview software.

7- System Design Considerations

The UNH prototype was built to use pumps as the method of mixing, instead of a stirrer mixer mounted through the tank with a mechanical seal. Mechanical seals are prone to leakage, which is particularly undesirable where volatile vapors are present as in biodiesel processing. The pump mixing system helps to improve overall safety of the processor.

Vent tubing is connected to all of the vessels such that the processor is a closed system other than the vent outlets. The vent tubing prevents siphoning and ensures no over pressuring. All stages of the system are operated at atmospheric pressure. For additional safety measures, the prototype processor is designed and built to fit inside the enclosure of a laboratory fume hood.

Pumps are expensive. The processor uses just two pumps for the whole system to minimize pumping capital costs. In order for two pumps to serve all of the mixing and fluid transfer purposes of the system, three way valves are used to direct piping streams. The system is designed to take advantage of gravity draining where possible.

The tubing diameter is chosen to slip over the glassware fitting ends. The pumping speed is designed to match the flow regime of the tubing. The fluid viscosity of oil is much greater than water and will of course affect the pumping rate. Once the oil is warmed and when reacted, the viscosity will drop and pumping performance improves. Valve options with the largest possible valve coefficient (Cv) values are specified. The three way valves used have a Cv of 0.17.

The primary reactor system is fitted with a heat cord apparatus fastened around the primary reactor to heat the oil and reaction mixture. The heat output of the heater therefore has to be limited in order to not fatigue the plastic fittings.

The methanol is reacted with catalyst to form methoxide solution. The alkaline catalyst used is potassium hydroxide (KOH) or sodium hydroxide (NaOH). Both catalysts come in crystalline flake form. The flakes dissolve in methanol solution and react to form methoxide. The methanol pump will not tolerate particulate through the impeller. To use the pump to mix the solution and dissolve the catalyst flakes, a screen catalyst trap is used at the top of the methoxide reactor. The methanol solution is pumped from the bottom of the methoxide reactor and into the top of the reactor. The catalyst trap is fitted to the end of the inlet tubing at the top of the reactor. The methanol flows through the trap, dissolving the catalyst while preventing any larger catalyst particulate from getting to the pump.

7- Automation Hardware

The sensing hardware are pressure sensors and thermocouples. The pressure sensors fitted to the bottom of the vessels are calibrated to display the fluid level in the vessels. To get an accurate pressure measurement, the system had to be both clear of bubbles and not circulating. The thermocouples used are type J thermocouples with a grounded pipe plug probe.

All of the valves are controlled by a Metrabyte mechanical relay board, with 24 relays. The relay board accepts digital 5 volt DC inputs from the DAQ and turns 120 volt AC power. The pumps and heater are controlled by solid state relays (SSRs). The SSRs switch 120 volt AC power with a 5 volt DC input. The advantage of SSRs is that they exhibit bounce-less operation which can be a concern when switch high power loads such as motor starters. The SSRs also features a larger heat sink for higher current load applications such as a heater. The variables measured, sensor and their use are indicated in the table below.

Variable Measured	How	Purpose/use
Pressure at bottom of reactor vessel	**Pressure transducer**	**Get/Display liquid level in tank Determine oil filling/product draining quantities Determine when Glycerin has been drained**
Pressure at bottom of Methoxide Tank	**Pressure transducer**	**Get/Display liquid level in tank Get methanol filling/ methoxide draining quantities**
Temperature..	**Type J thermocouple**	**- Read temperature measurement through DAQ in Labview**
Pump flow	**Pressure transducer**	**- Read Pressure head through DAQ in Labview - Calibrate program to show flow rate**

The Parameters controlled and the functions in the processing are indicated in the tables below.

Parameter Controlled	How/Purpose
Reactor filling and draining period	**The liquid level (or liquid volume) in the tank, as measured by the pressure transducer.**
Methanol filling and draining period	**The liquid level (or liquid volume) in the tank, as measured by the pressure transducer.**
Mixing times	**Dissolving/reacting/soap and methanol removal, as measured by labview timer.**
Temperature	**Reaction temperature and methanol removal temperature.**

7- Automation Software

Control logic is performed in PC software Labview. The computer is interfaced with the hardware components by the USB DAQ and modules. Labview was used to develop the automation program (see picture of front panel). Labview reads analogue voltage signals from the pressure sensors and thermocouples modules. The outputs are to a 32 channel sourcing module linked with the relay board. The Labview program continually monitors pressure, temperature and time; in addition to responding to user inputs.

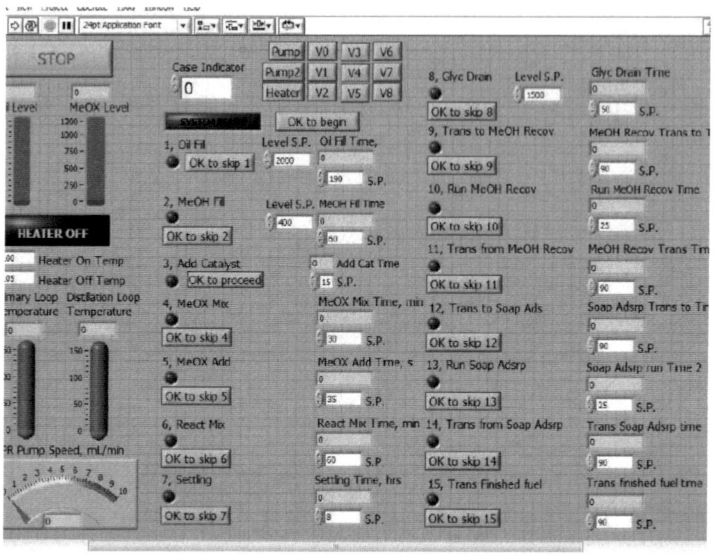

8- Challenges /Difficulties

One of the most significant challenges of the project was to find appropriate materials that are compatible with the aggressive nature of the chemicals being used. For observational purposes the system was to have clear tubing that is also flexible to be easily fitted to laboratory glassware. The tubing also had to be able to withstand high temperatures encountered in the reacting and in particular the methanol recovery stages. After extensive research a Tygon formulation was decided on, based on the manufacturer's recommendations. However, in experimentation the tubing did not perform adequately. The solvent nature of the biodiesel, in combination with the concentrated alcohol and extremely alkaline catalyst, fatigued the tubing over time and developed leaks at the fittings. A future design should use a thicker walled tubing and if possible more chemical resistant polymer. The pressure sensors used are an expensive component. Much research was done to find sensors that could operate with the desired sensitivity at a reasonable price. Sensors that were made from chemical resistant materials such as stainless steel were overly expensive. Brass sensors were purchased and installed with oil filled tubing traps to isolate the sensors from the alkaline catalyst. All other components of the system were made of Pyrex glass, stainless steel or high density polyethylene.

Measurement sensitivity was a significant challenge for the pressure sensors. These are expensive and the price increases exponentially with the degree of sensitivity of the sensor. Pressure sensors for the processor are used to measure the height of fluid in the vessels. The sensors were calibrated in Labview to return the volume of fluid filling each vessel with±10mL. The operation of the sensors was sometimes erratic. In order for the sensor to give an accurate reading the fluid in the vessel had to be both stagnant and the lines containing the pressure sensor free of air bubbles.

Fluid resistances of the plumbing system caused significant back pressure on the pump and a slower than desired pumping rate. The achieved flow rate was about 70% of the desired flow rate. To compensate the reaction fluid was circulated for a longer period of time.

The Amberlite adsorbent used is a small particle polymer resin, and was placed in a fluidized bed. The bed had to accommodate the low fluidization velocity of the particles.

9- Results

A benchtop prototype automated system has been designed and built. It simulates typical industrial scale processing system, with automation. An automation loop is implemented using Labview and National Instruments hardware to control the core operations of the processor. The results of the system lay the ground work for design of an automated industrial scale processor.

Acknowledgment:

The authors would like to acknowledge the support of the NH Industrial Research Council, The UNH Hamel Center for Undergraduate Research, The University of New Hampshire Instrumentation Center, and Micro Biofuel Processors, LLC.

Works Cited

Benefits of Biodiesel. National Biodiesel Board. 6 August 2008 <http://www.biodiesel.org/pdf_files/fuelfactsheets/Benefits%20of%20Biodiesel.Pdf>

Bio Energy. Agroconsult. 6 August 2008 <http://agroconsult.netfirms.com/>

Clean Air USA. Clean Air USA. 3 August 2008 <www.cleanairusa.org>

Demirbas, Ayhan. Biodiesel: A Realistic Fuel Alternative for Diesel Engines. New York: Springer, 2008.

Emissions. National Biodiesel Board. EPA. 6 August 2008 <http://www.biodiesel.org/pdf_files/fuelfactsheets/Benefits%20of%20Biodiesel.Pdf>

Pahl, Greg. Biodiesel: Growing a New Energy Economy. Vermont: Chelsea Green Publishing, 2007.

394h The Future of Biodiesel as a Viable Alternative Fuel: Challenges and Opportunities

K. Y. Simon Ng[1], Steven O. Salley, John Wilson[2], Haiying Tang[1], Shuli Yan, Nadia J. Abunasser[3], and Manhoe Kim[4]. (1) Department of Chemical Engineering and Materials Science, Wayne State University, 5050 Anthony Wayne Dr., Detroit, MI 48202, (2) National Biofuel Energy Laboratory, NextEnergy, 461 Burroughs ST, Detroit, MI 48202, (3) Chemical Engineering and Materials Science, Wayne State University, 461 Burroughs, Detroit, MI 48202, (4) National Biofuel Energy Laboratory, NextEnergy, USA, 461 Burroughs Street, Detroit, MI 48202

Biodiesel is experiencing the growing pain as a sustainable biofuel. The "fuel vs. food" debate has an unfavorable impact on the biodiesel industry, since majority of biodiesel is produced using refined vegetable oil. In fact, the cost of soybean oil has increased significantly in the past year. This presentation will examine a number of technical, societal, and economic challenges facing the biodiesel industry. Opportunities to develop a truly sustainable and renewable diesel will be discussed.

Electrical Characterization of Lead-Acid Battery at Load for HEV Applications

*Wenhua H. Zhu, Ying Zhu, Robert U. Payne, and Bruce J. Tatarchuk**

Center for Microfibrous Materials Manufacturing
Department of Chemical Engineering
230 Ross Hall, Auburn University, AL 36849
brucet@eng.auburn.edu, zhuwenh@auburn.edu

Lead-acid batteries have been widely used as secondary batteries for more than a hundred years. For cost consideration, it is still a good resource for fast starting of cold vehicles, for recharging from either a stop-start braking system, or for a charge from the engine itself, which consumes battery energy or stores electricity back into chemical energy. Battery modeling is a key issue for different battery applications. It is necessary to develop a reliable and easy-to-parameterize battery model. In this work, the impedance technique is applied to dynamic modeling of battery behavior and diagnosis of quantities like actual capacity, state of charge (SoC) and state-of-health (SoH). Interpretation of the test results and proper modeling requires that the experimental setup allows some idealization like separating between working electrodes with reference electrodes or maintaining constant concentrations of reactants at the interface. The measurement and interpretation of the impedance of lead-acid batteries is further discussed, especially on nonlinearity, voltage drift, stability, half-cell measurements, model structure, and parameter extraction from the impedance data. This initial work focuses on the methods and procedures for testing lead-acid batteries at load and development of a reliable circuit model for predicting battery performance, which provides an opportunity to simulate a sophisticated lead-acid battery power system for HEV applications in a real time environment.

***Keywords**: Lead-acid battery; AC impedance; equivalent circuit; electrical characterization.*

The advantages of the lead acid system are its high-rate discharge capability, good specific energy, high reliability, robustness, and low-cost in both manufacturing and recycling, as the battery is manufactured mainly from a single low-cost raw material.[1, 2] Electrochemical impedance spectroscopy (EIS) is an excellent tool to analyze the interfacial process, variation in the internal resistance, state-of-charge, and the residual capacity of a lead-acid battery. *Ac* impedance is a fast, accurate, and non-destructive in nature for the modeling and diagnosis of industrial batteries.[3-5] The basic experiment of the EIS method consists of applying a small, sinusoidal voltage or current signal to an electrochemical cell, measuring the system's response with respect to amplitude and phase, and determining the impedance of the system over a spectrum of frequencies. Interpretation of the results and proper modeling requires that the experimental setup allows some idealization like separating between working electrodes and reference electrodes or maintaining constant concentrations of reactants at the interface.[6] The impedance technique can be applied to dynamic modeling of battery behavior and diagnosis of quantities like actual capacity, state of charge (SoC) and state-of-health (SoH).[7] The SoH influences mostly the ohmic resistance, the charge transfer resistance R_{ct} and the parameters of the constant phase element accounting the diffusion in the pores when the plate is well charged. Battery modeling is a key issue for design of different battery applications. It is necessary to develop a reliable and easy-to-parameterized battery model. However, for the lead acid battery, the impedance spectra depend very much on the short-term charge-discharge history. Karden *et al.*[7] further discussed methodological questions connected with the measurement and inter-

*Corresponding author. Tel.: +1-334-844-2023.

pretation of the impedance of batteries, especially nonlinearity, voltage drift, stability, reproducibility, half-cell measurements, model structure, and parameter extraction from the impedance data. This initial work uses a 12 V auto lead-acid battery and ac impedance methods to collect the real time data at load conditions. Through the development of a suitable equivalent circuit to simulate the measured impedance data and predict reliable battery performance, the obtained parameters and physical elements can be further applied into PSpcie simulation or dynamic modeling of the HEV power system.

In this work, an EG&G Model 273A/ 5210 impedance system was first applied to the test of a defective lead-acid battery. The battery had no power left inside and could not be charged back to the normal operating condition. The defective battery voltage (several volts or mini-volts) are located in the normal 10 V range of the instrument requirement for impedance test. Secondly, using a Gamry FC350™/TDI electronic load, a 12 V Duralast 51-D lead-acid battery (a healthy battery with a capacity of 85 Ah, cold cranking of 500 Amp) was charged overnight at a current of less than 500 mA with a voltage no more than 12.8 V. The FC350 impedance system, working in hybrid impedance mode, modulates the current from the battery at load. Simultaneously, the current information at the electronic load is sent to the FC350™ monitor. The battery voltage is measured by the FC350 directly. The FC350 system collects these testing data and generates the impedance data files. The Hybrid EIS mode was applied for the experiments in order to observe the EIS behavior at low frequencies. Better results on ultra-low impedance analysis at a high power output can be obtained using this impedance system.

A simplified equivalent circuit R (RC_{pe}) for the defective lead-acid battery was applied to the NLLS fitting process (Figure 1). The three values of physical elements, *i.e.* R_s—uncompensated resistance related to the membrane/ solution, R_p —charge transfer resistance related to the polarization process, and C_{pe}—constant phase element related to electrode surface area or electrode roughness with a distribution of active sites on the surface. In another word, this physical element is a replacement of the double layer capacitor in

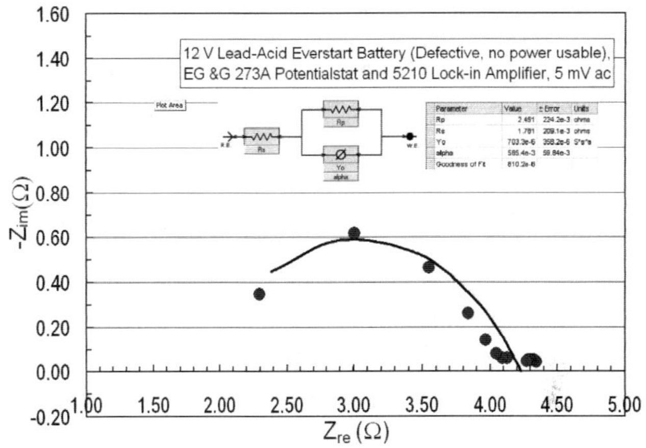

Figure 1. Nyquist plot for a defective Everstart lead-acid battery without a load. Data at high frequency side was not available using this impedance system.

considering the heterogeneous reaction rates on an electrode surface. For the specified defective lead-acid battery, a small amount of charge changes the shape of the Nyquist curve due to the change in resistance of the battery. The impedance test on the defective battery was conducted in order to understand the degradation mechanism. Extensive test for the performance and capacity degradation will be conducted in the future work.

Figure 2 shows the Nyquist plot of a normal healthy lead-acid battery which was tested at 1.0 A load via ac impedance. The battery voltage of 12.3 V before the test, charged overnight at a current of less than 500 mA with a voltage no more than 12.8 V. The assumed equivalent circuit model is written as $LR(RC_{pe})(RC_{pe})$. Impedance data is well fitted to the assumed circuit model. The errors near to the center area of the Nyquist semicircle were not clear. It appears to be related to the chemical crystallization and dissolution processes.[8] Further work will focus on the analysis of the impedance data at a higher current load and development of the circuit model related to the physical process, chemical and electrochemical mechanism.

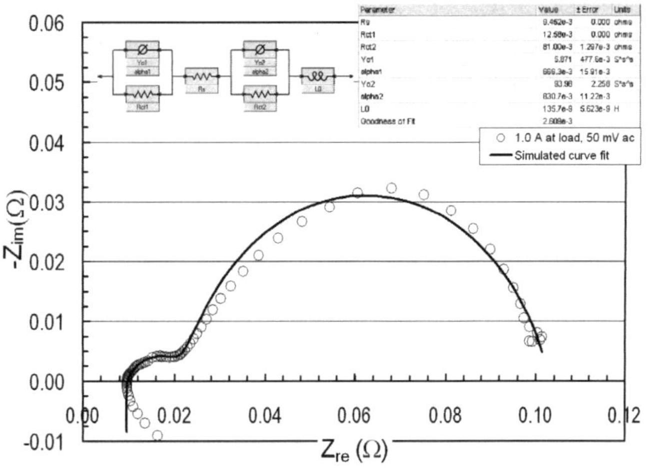

Figure 2. Nyquist plot for testing of a 12 V Duralast 51-D lead-acid battery at load of 1.0 A current (capacity 85 Ah, cold cranking 500 A). Battery voltage 12.3 V before test, charged overnight at a current of less than 500 mA with a voltage no more than 12.8 V.

From AC impedance data collected at different currents, results are analyzed and compared between the circuit elements and physical processes determined via AC impedance and independent measurements of these same elements. Through different loadings (galvanostatic method), the more understanding about the mechanism and circuit model for the lead-acid battery will be obtained in the future work. The physical elements at load can be applied to PSpice circuit simulation or to test the dynamic load for HEV applications. The real time evaluation of lead-acid batteries potentially gives better understanding of the physical process, chemical reaction, and electrochemical mechanism in the dynamic power system.

References

1. Hejabi, M.; Oweisi, A.; Gharib, N., Modeling of kinetic behavior of the lead dioxide electrode in a lead-acid battery by means of electrochemical impedance spectroscopy. *Journal of Power Sources* **2006,** 158, (2), 944-948.
2. Parker, C. D., *J. Power Sources* **2001,** 100, 18-28.
3. Huet, F., A review of impedance measurements for determination of the state-of-charge or state-of-health of secondary batteries. *Journal of Power Sources* **1998,** 70, (1), 59-69.
4. Salkind, A.; Atwater, T.; Singh, P.; Nelatury, S.; Damodar, S.; Fennie C, Jr.; Reisner, D. In *Dynamic characterization of small lead-acid cells*, Manchester, 2001; Elsevier Science B.V.: Manchester, 2001; pp 151-159.
5. Salkind, A. J.; Singh, P.; Cannone, A.; Atwater, T.; Wang, X.; Reisner, D., Impedance modeling of intermediate size lead-acid batteries. *Journal of Power Sources* **2003,** 116, (1-2), 174-184.
6. Macdonald, J. R., *Impedance Spectroscopy: Emphasizing Solid Materials and Systems.* Wiley-Interscience: 1987.
7. Karden, E.; Buller, S.; De Doncker, R. W., A method for measurement and interpretation of impedance spectra for industrial batteries. *Journal of Power Sources* **2000,** 85, 72-78.
8. Thele M.; Karden E.; Surewaard E.; Sauer D.U., Impedance-based overcharging and gassing model for VRLA/AGM batteries, *Journal of Power Sources* **2006,** 158, 953-963.

457b Addressing Challenges In the Simulation of Lithium-Ion Battery Models

Vijayasekaran Boovaragavan and **Venkat Subramanian**. Chemical Engineering, Tennessee Technological University, 1020 Stadium Drive, Prescott Hall Room 307, Cookeville, TN 38505

Though Lithium-Ion battery was commercialized in 1991, it has several challenging issues to be addressed for potential applications like satellite, military and hybrid vehicles.1-6

Usually, the lithium-ion battery models have multiple partial differential equations in multiple dimensions (namely spatial coordinate x along the electrode and spatial coordinate r along the solid particles) with highly nonlinear kinetic and transport expressions that govern the entire electrochemical behavior of the battery system. Hence, with this nature of the model, its simulation process is computationally challenging even for normal design and analysis purposes. But, real-time simulation is necessary due to the technological developments in power systems that includes hybrid and stand alone environments.

With growing power requirements, batteries and fuel cell are playing vital role in advanced technologies like a portable power system that can power the entire building or office spaces. This also necessitates parameter estimation and capacity fade analysis in battery applications.

Recent researches7-12 that support the real-time simulation and parameter estimation revealed that advanced mathematical methods with thorough knowledge in battery system is needed to convert the highly nonlinear and multi-dimension battery models in to a mathematical form that can govern the battery behavior without losing physical significance.

This talk will present these and other developments in the real-time simulation and parameter estimation using a rigorous lithium-ion battery model. Progresses made in simulating different discrete events in battery management, cycling studies will also be presented. For example, a code that simulates 1000 cycles of charge-discharge behavior of a typical lithium-ion cell using physics based rigorous models in less than minute has been developed and will be presented.

Acknowledgements

The authors would like to acknowledge the US Government, US Army CECOM and NSF for funding various lithium-ion batteries modeling related work. The corresponding author VS would like to acknowledge Professor John Newman, University of California Berkeley for suggesting the need for real-time simulation of lithium-ion battery models.

References

1. G. G. Botte, V. R. Subramanian, and R. E. White, 2000, Electrochim. Acta, 45, 2595.

2. V. R. Subramanian, V. D. Diwakar, D. Tapriyal, J. Electrochem. Soc., 152(10), A2002-A2008 (2005).

3. V. D. Diwakar and V. R. Subramanian, J. Electrochem. Soc., 152(5), A984-A988 (2005).

4. S. Devan, V. R. Subramanian and R. E. White, J. Electrochem. Soc., 152(5), A947-A955 (2005).

5. Q. Guo, V. R. Subramanian, J. W. Weidner and R. E. White, J. Electrochem. Soc., 149(3), A307-A318 (2002).

6. V. R. Subramanian, P. Yu, B. N. Popov and R. E. White, J. Power Sources, 96(2), 396-405 (2001).

7. Vijayasekaran Boovaragavan and Venkat Subramanian. Journal of Power Sources, 173(2), 1006-1011, 2007.

8. Venkat Subramanian, Vijayasekaran Boovaragavan and Vinten Diwakar. Electrochemical and Solid-State Letters, 10(11), A225-A260, 2007.

9. Vijayasekaran Boovaragavan and Venkat Subramanian. Electrochemistry Communications, 9(7), 1772-1777, 2007.

10. Venkat Subramanian, Vijayasekaran Boovaragavan, Kartik Potukuchi, Vinten Diwakar and Anupama Guduru. Electrochemical and Solid-State Letters, 10(2), A25-A28, 2007.

11. Vijayasekaran Boovaragavan, Vinten Diwakar and Venkat Subramanian, "Review of various simulation approaches for lithium-ion battery models" in "Advanced Materials and Methods for Lithium-Ion Batteries," Research Signpost, India, in press, 2007. (Invited article)

12. V. R. Subramanian, V. Boovaravan and V. D. Diwakar, "System Level Component Models for Electrochemical Power Sources in Hybrid Environments", The 2006 Annual AIChE meeting, San Francisco, November 2006.

Robust Surrogate Modeling of Lithium Ion Batteries

Rajeswari Chandrasekaran[1, 2], Jonathan Denis[1], Michael Skinner[1], Thomas F.Fuller[1, 2]

(1) Center for Innovative Fuel Cell & Battery Technologies, Georgia Institute of Technology, Atlanta, GA 30332, (2)School of Chemical & Biomolecular Engineering, Georgia Institute of Technology, Atlanta, GA 30332.

Introduction

Significant research is being done in the field of lithium ion batteries for hybrid electric vehicles and plug-in hybrid vehicles by different automakers. Various physics based models exist in literature that describe in detail the initial performance and also the cycle life or storage issues. But the knowledge from physics based models with respect to both performance and durability should be incorporated in the vehicle level system model earlier in the design phase. The emphasis on this approach is increasing as the paradigm is shifting to design for affordability. Incorporating physics based models in system model is complicated due to different languages in which they are coded and also because of the slower computation speed of the physics based models as numerical methods are involved. Hence a systematic framework is necessary to approach this problem. Robust design through surrogate model methodology is useful in this regard to develop simpler models without losing much of the fidelity and they are easy to integrate in vehicle model. Response surface methodology is a key enabler for surrogate models. This methodology allows one to explore the design space for multiple responses of interest –capacity, power, specific energy, cycle life, etc. and determine the optimal settings by means of a desirability function approach. Overall desirability can be obtained from the individual desirability functions of the responses. This means a trade-off between multi- objectives is obtained. The robust design provided by the surrogate model methodology is also less sensitive to uncertainties or noise factors. Both physics based models or experimental data can be used to arrive at response models. Experiments avert the need for assumptions associated with physics based models, whereas the latter is needed for cases where historical or experimental data are unavailable.

In the current research, the surrogate model approach is introduced for analyzing performance of lithium ion batteries. This can be extended in the future to include cycle and calendar life as well to study degradation of batteries at system level.

Response Surface Methodology (RSM)

Even though the system models that are available currently have battery models as one of the components, most of the battery models use look up tables for OCP or SOC values. The disadvantage of look up tables is that they have limited scope in terms of data available, e.g., OCP data may not be available at different temperatures. Another disadvantage is that the interpolation procedure used can lead to unknown errors. Also, determination of SOC of a battery onboard a vehicle is difficult and is a source of uncertainty. Simple battery models in current system models may be sensitive to such uncertainties. Hence, there is need for a surrogate model based on RSM which facilitates a robust design space solution insensitive to uncertainties. Discussion on RSM can be found elsewhere (1). The simple response surface equation (RSE) is a quadratic equation based on Taylor series approximation.

Application of RSM for battery surrogate model

Lithium ion cell from Valence technology with rated capacity of 1.1 Ah (C rate) is chosen for the experiments. The controllable factors chosen were temperature of operation and rate of discharge. It should be noted as the experiment proceeds, there will be change in temperature of the battery, especially at high rates of discharge and there will definitely be a temperature gradient within the battery. In this way, the temperature of the battery is a noise factor (or uncontrollable factor). But for the time being, constant temperature of operation of the battery is considered. Once a good RSE is developed, one can add random distribution to the output to reflect these noise factors during analysis. For each of the independent variable, three levels are chosen. A central composite design of experiments is created and total of 9 runs were conducted using an Arbin battery cycler (BT-2000) and HD-508 environmental chamber from Associated Environmental Systems. The design of experiments is shown in Table 1. The CCD design gives the same number of runs as a full factorial in this case since there are only two independent variables. All rates of discharge are denoted in reference to C rate.

Temperature (°C)	Rate of Discharge (C rates)
0	0.5
0	1.75
0	3
25	0.5
25	1.75
25	3
50	0.5
50	1.75
50	3

Table1. CCD Design of Experiments

Results and Discussions

The discharge curves for the above 9 cases for Valence cell are given in Figure 1.

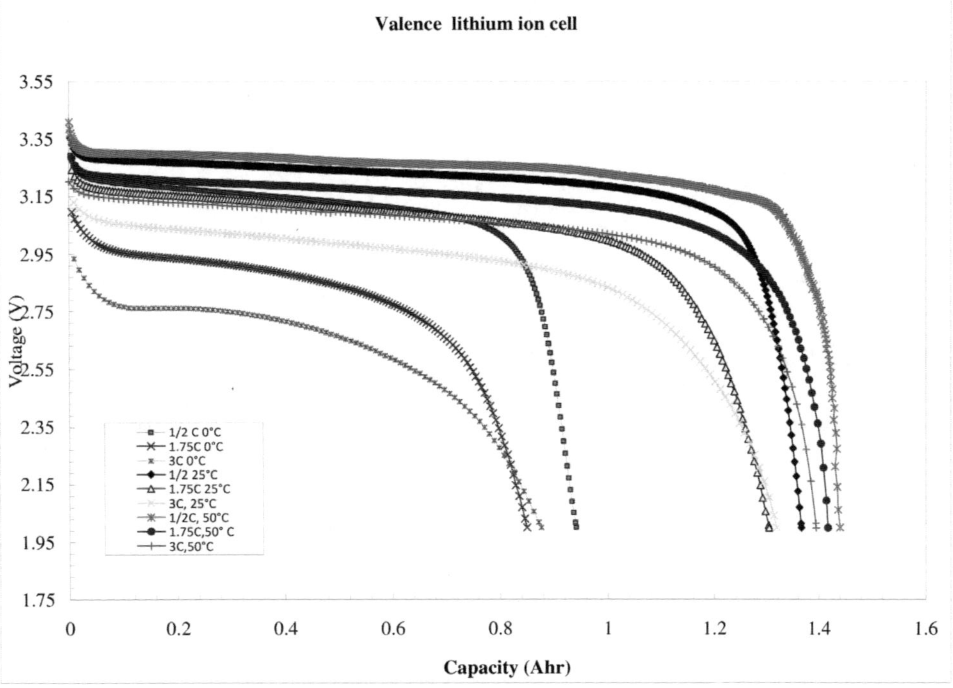

Figure 1 Discharge curve of Valence 1.1 Ah lithium ion cell

Response Surfaces were created for the capacity obtained in the above runs as a function of temperature and rates of discharge. The results are now analyzed using statistical goodness of fit tests.

Actual vs. Predicted Plot

Figure 2 gives the actual vs. predicted plot for the capacity (Ah). The R^2 value is 0.99. All data points lie close to the Perfect Fit line (diagonal). But they don't all seem to be distributed evenly and this might occur when a single variable is the dominant factor. The mean of the response is also slightly shifted upwards because of the uneven distribution. More cases could be run to check if this observation still exists.

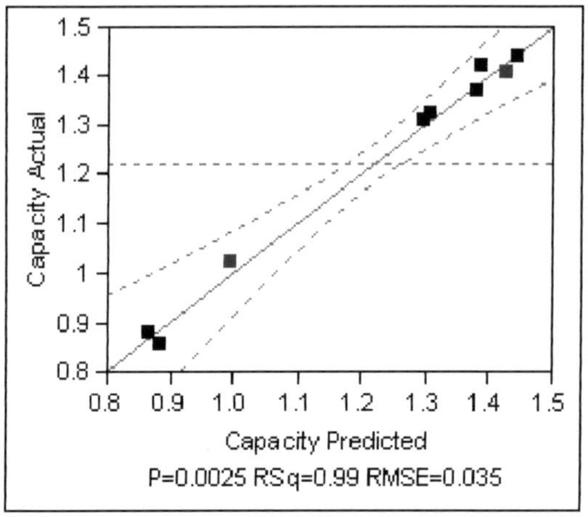

Figure 2 Actual vs. Predicted plot of capacities (Ah) of Valence lithium ion cell

Residual vs. Predicted Plot

Figure 3 gives the residual vs. predicted plot. There is no distinguishable pattern observed in the residual by predicted plot. A good random distribution of the error would imply that it was fine to have discarded the higher order terms and interactions from the Taylor series expansion for the assumed model. But slight data clumping towards right corner is seen. This could again imply that single variable is driving the response in the regions of interest. The total span of error is 10% of the minimum of the predicted capacity values. Usually it is desirable to have this as low as 5% as a rule of thumb for validity of the second order meta model.

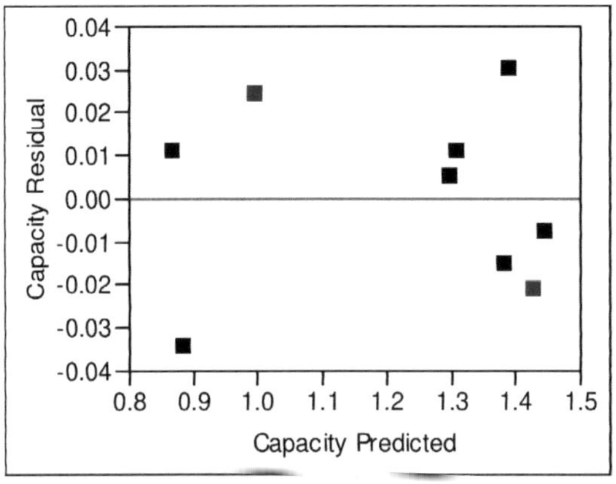

Figure 3 Residual vs. Predicted plot of capacities (Ah) of Valence lithium ion cell

Model Fit Error (MFE)

This statistic gives an idea of how well the model fits the data points in the Design of Experiments. Ideally, it is desired that the error distribution resemble a normal distribution with mean zero and standard deviation of one. Figure 4 gives the model fit error for the predicted capacities.

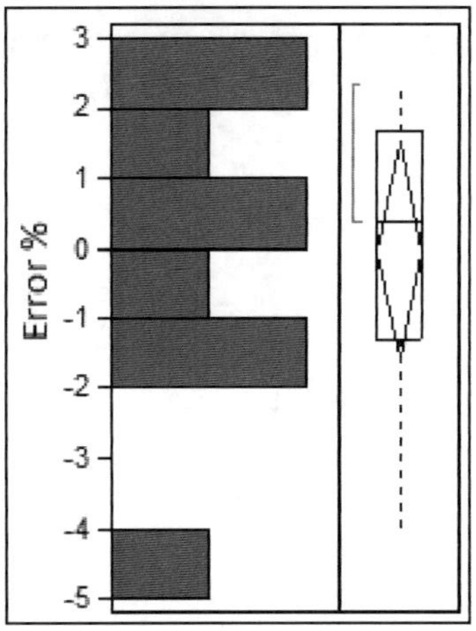

Figure 4 Model Fit Error Distribution

As can be seen in Figure 4, the upper and lower bounds of the error are less than 5%. Usually, it is desirable to minimize the bounds. The mean and standard deviation of the above distribution are 0.051 and 2.01 respectively. It is desirable to have mean approximately equal to zero (as obtained) and standard deviation less than one. In the above distribution, if the outlier is neglected, then the standard deviation reduces to 1.43 and the error bounds are also minimized to 2.3 and -1.3 % respectively. But caution has to be applied while neglecting outliers to avoid significant correlation between the independent variables. Usually, it is okay to neglect about 7-8% of the outliers. Since in this work there were only nine cases to begin with, neglecting of outliers is avoided. The correlation between the independent variables with all 9 runs is zero. It is necessary to try and fix the shortcomings in the assumed model until this stage and get the goodness of fit statistics better before proceeding further. But as a learning exercise, more in this methodology is presented here. A point to be noted above is that there were fewer runs and hence normal distribution of the error within this small data set might be slightly difficult.

Model Representation Error (MRE)

This statistic shows how well the assumed model predicts the actual response for the design settings not used in the creation of model in the entire design range of interest. For this, random settings of the independent variable are chosen and experiments performed. A subset of a Latin hypercube sample was used to arrive at some of the cases below. Usually it is desired to have at least 20% of the original number of runs to have a sufficient sampling of the space to check the MRE. The RSE developed above is used to predict the responses and arrive at MRE by comparing it to the actual capacities.

Temperature (°C)	Rate of Discharge (C rates)
6	1.07
11	2.03
18	3
22	0.74
50	0.5
0	1
0	2
25	1
25	2

Table 2 Random data for Model Representation Error Distribution

Figure 5 shows the MRE distribution. It can be seen that the lower error bound has increased as well as the mean (-2.08) and standard deviation (3.63). Even though it is understandable to see an MRE with poorer statistics than MFE, a good RSE will lead to a better MRE. It should also be noted that if the data at zero degrees are removed from the above table, the mean and standard deviation of the MRE distribution as well as the lower bound significantly reduce. This implies that the predictive capability of the assumed model reduces at 0°C.

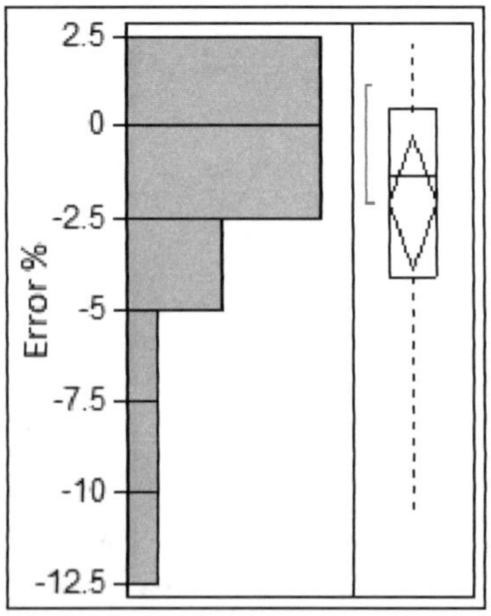

Figure5 Model Representation Error Distribution

Latin hypercube (LH) sampling is basically a space filling design and complements the central composite design. In order to improve the assumed model, a space filling design could be used alongside CCD to create the RSEs. As a first attempt, the extra cases in Table 2 were used alongside Table 1 to create the RSEs and the goodness of fit statistics was observed. The statistics didn't improve much. So higher order terms could be employed or a proper LH design could be employed to arrive at a rigorous RSE.

Response Surface Equation (RSE)

Once a good RSE is obtained, it represents the meta model of the metrics considered as a function of the variables. Figure 6 shows the response surface of capacity as a function of rate of discharge and temperature.

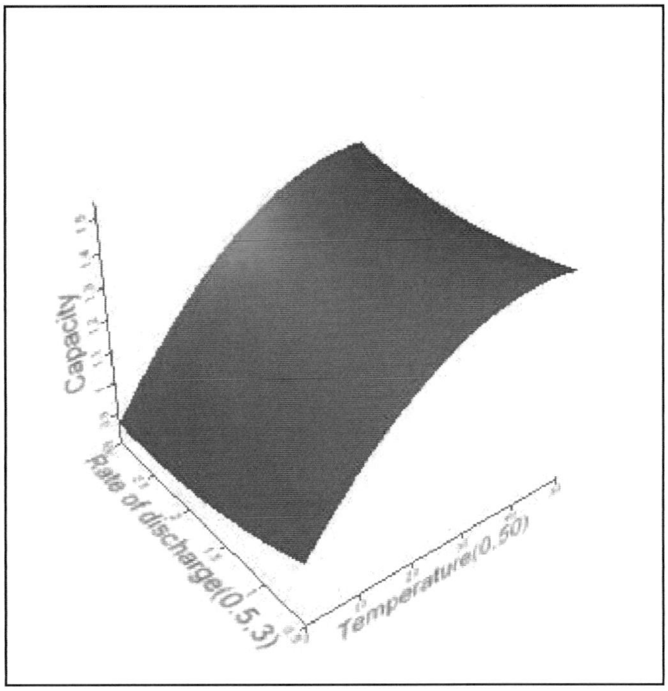

Figure 6 Response Surface of capacity of 1.1 Ah Valence lithium ion cell

Analysis of influence of the independent variables

Figure 7 shows the Pareto plot. This plot gives the relative influence of the variables and the interaction between variables on the response, i.e. capacity. It can be seen that temperature has a significant effect on the capacity.

Term	t Ratio
Temperature(0,50)	17.63075
Temperature*Temperature	-6.51896
Rate of discharge(0.5,3)	-2.61147
Rate of discharge*Rate of discharge	1.89793
Temperature*Rate of discharge	1.61650

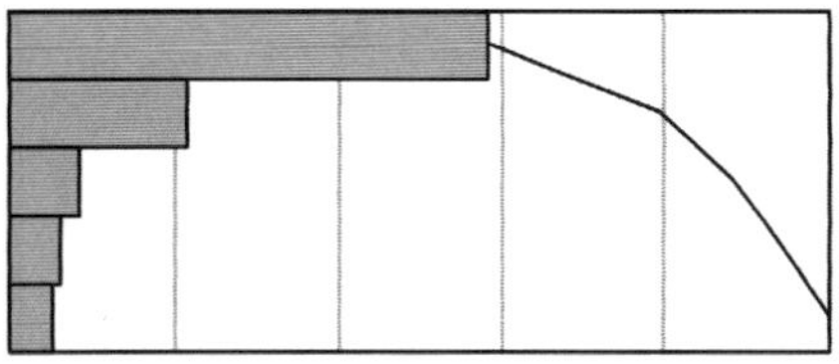

Figure 7 Pareto plot

The influence of the individual variables on the response can be viewed either in Figure 8 in a prediction profiler or in Figure 9 in the Scatter plot matrix. This gives the change in capacity of the cell with respect to temperature or rate of discharge alone while the other variable remains fixed at a value within the range of interest. One can then obtain simulated responses using the RSEs created for, say 5000 runs, by using a random uniform distribution on temperature as well as rate of discharge within the design range. A random noise can also be added on the capacity values. In this case random noise with standard deviation of 0.0349 is added to the responses. The resulting capacity distribution is shown on the right most plot in Figure 8. The corresponding scatter plot matrix is given in Figure 9.

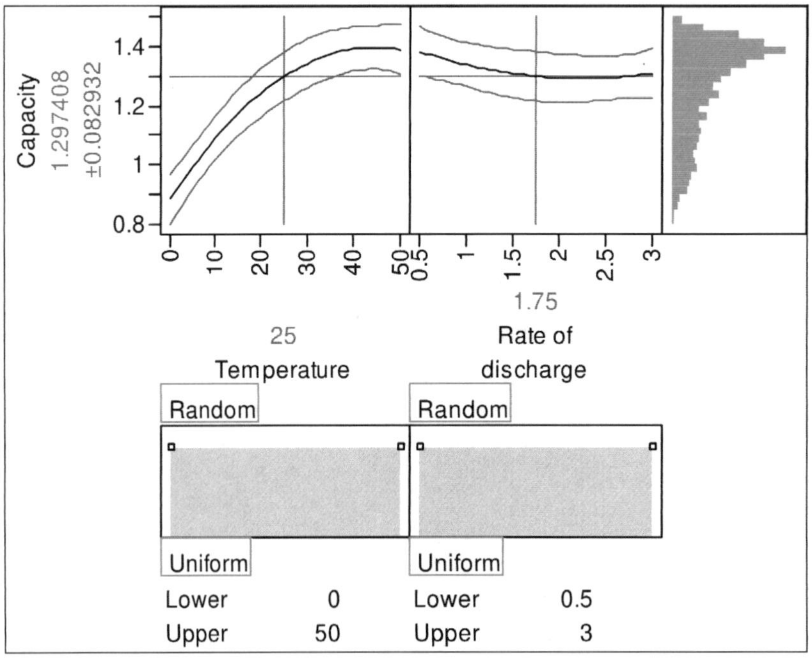

Figure 8 Prediction Profiler

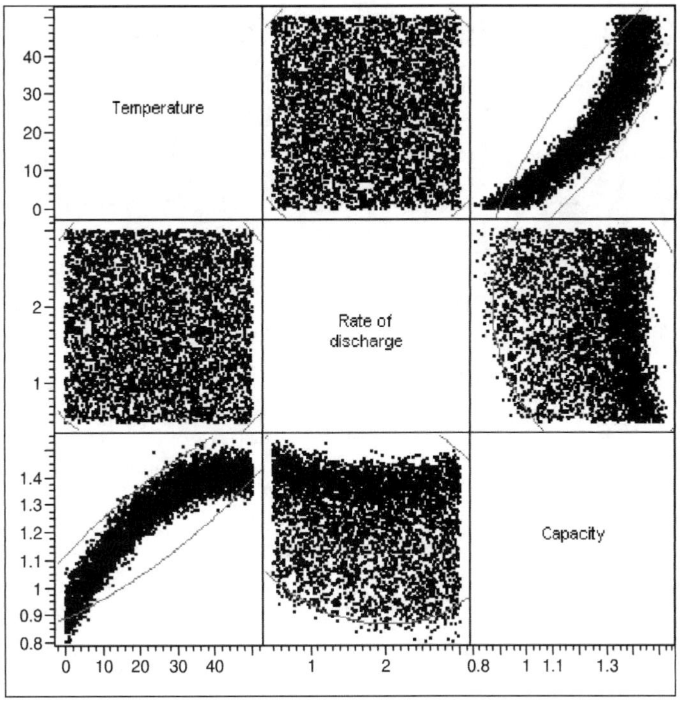

Figure 9 Scatter Plot matrix

Desirability functions can also be used to arrive at optimal settings of operation to achieve the objective.

Suggestions for future work

The RSEs when used outside the design range, e.g., at -25^{0}C or at rates lower than 0.5C (e.g. 0.1C) or greater than 3C (e.g. 4C), give greater error. Hence in order to obtain good RSEs, one might have to expand the design range. At the same time, if the design space is expanded too much, the central composite design of experiments alone will not be sufficient to obtain good RSEs. It has to be coupled with one of the space filling designs. Several other performance metrics like power, specific energy can also be studied and multiple RSEs can be created. This methodology can also be extended to understand the influence of variables like electrode thickness, porosity, etc. on the capacity by using physics based models to obtain the responses that can be used to develop response surface equation. Finally, the RSM can be extended to study degradation in batteries by having number of cycles or storage time as one of the independent variables of interest. The interaction of temperature with cycle life and calendar life on capacity will be of significance also.

Acknowledgements

Georgia Tech Research Institute is gratefully acknowledged for financial support.

Reference

(1) Rajeswari Chandrasekaran, Wu Bi and Thomas F. Fuller, *J.Power Sources, 182(2), Aug 2008, 546-557.*

457d Impact of Ultra-High Carbon Surface Areas on Capacitory and Battery Performance

Galen J. Suppes and Bryan Sawyer. University of Missouri-Columbia, Department of Chemical Engineering, W2028 Engineering Bldg. East, Columbia, MO 65211

Activated carbon is widely used in battery and supercapacitor applications. Surface area is a critical property that leads to improved power density and, in some instances, energy density. Activated carbons are generally available with BET surface areas less than 2,000 m2/g. This proposal is on the use of ALL-CRAFT carbons with BET surface areas in excess of 3,000 m2/g for several capacitor and battery architectures.

457e Silicon Nanotube. A Promising Candidate for Hydrogen Storage

Dapeng Cao, Jianhui Lan, and Wenchuan Wang. College of Chemical Engineering, Beijing University of Chemical Technology, Bei-San-Huan East Road 15, Beijing, 100029, China

A multiscale theoretical method, which combines the first-principle calculation and a grand canonical Monte Carlo (GCMC) simulation, is used to predict adsorption capacity of hydrogen in silicon nanotube (SiNT) arrays at T=298 K in the pressure ranges from 1 to 10 MPa. In the multiscale method, the binding energy obtained from the first-principle calculation is used as an input in the GCMC simulation. It is found from the first-principle calculation that the SiNT arrays exhibit much stronger attraction to hydrogen both inside and outside SiNTs, compared to the isodiameter carbon nanotubes (CNT). The subsequent GCMC simulations indicate that the SiNT arrays present distinct improvements of 106 %, 65 % and 52 % in gravimetric adsorption capacity of hydrogen at P=2, 6 and 10 MPa, respectively, compared to the isodiameter CNTs. It suggests that the SiNT array is a promising candidate for hydrogen storage.

457f Dynamics of Ti-Al-H Compounds In Ti-Doped $NaAlH_4$: A DFT-MD Stability and Transport Study

Gopi Krishna Phani Dathara and Daniela S. Mainardi. Louisiana Tech University, 911 Hergot St., Ruston, LA 71272

It is believed that the diffusion of atomic/molecular hydrogen and ionic species in metal hydrides is improved by the presence of dopants [1]. However, the role of transition metal dopants in improved thermodynamics and kinetics of hydrogen ab/desorption is still debated in solid state hydrogen storage research until date. Sodium aluminum Hydride ($NaAlH_4$) also known as sodium alanate, decomposes through a two step reaction to yield 5.6 wt% of hydrogen, which results in Na_3AlH_6, Aluminum and NaH phases [2]. Information regarding the formation of these phases provides insight into the rate limiting step for the hydrogen desorption process from metal hydrides [3]. Moreover, when Ti is present in the sodium alanate lattice, the stability and transport properties of the Ti-Al-H complexes that are formed would shed light into the role of Ti (dopant or catalyst or both) when added to sodium alanates.

In this work, Ti-dopant effects in the sodium alanate bulk, (001) surface, and on-top (001) surface are studied considering Na and interstitial lattice sites. The stability of the formed Ti-Al-H complexes and their dynamics over time at different temperatures are investigated using periodic Density Functional Molecular Dynamics (DFT-MD) simulations [4]. From calculations of the mean squared displacement (MSD) for different species, transport properties such as diffusion coefficients are calculated from the slope of the MSD curve. From these results, the role of Ti and its effect on the diffusion of atomic/ionic species and groups within the system can be elucidated.

[1] D. L. Anton, *Hydrogen Desorption Kinetics in Transition Metal Modified $NaAlH_4$*, J. Alloys Compd. 356-357 (2003), pp. 400-404.

[2] B. Bogdanovi• and M.Schwickardi, *Ti-doped alkali metal aluminium hydrides as potential novel reversible hydrogen storage materials*, J. Alloys Compd. 253 (1997), pp. 1-9.

[3] H. Gunaydin, K. N. Houk and V. Ozolins, *Vacancy-mediated dehydrogenation of sodium alanate*, Proc. Natl. Acad. Sci. U. S. A. 105 (2008), pp. 3673-3677.

[4] G.K.P. Dathara and D.S. Mainardi, *Structure and Dynamics of Ti-Al-H Compounds in Ti-Doped $NaAlH_4$*, Mol. Sim. 34 (2008), pp. 201-210.

457g New Insights into the Mechanism of Hydrolysis of Sodium Borohydride

Amy M. Beaird[1], **Eyma Y. Marrero-Alfonso**[1], Thomas A. Davis[2], and Michael A. Matthews[1]. (1) University of South Carolina, Department of Chemical Engineering, Columbia, SC 29208, (2) Department of Chemical Engineering, University of South Carolina, Columbia, SC 29208

Chemical hydrides have promise as a compact medium for hydrogen storage at low temperatures and pressures. Hydrogen is produced when a chemical hydride, such as NaBH4, exothermically reacts with water.

$$NaBH_4 + (2+x)\, H_2O \rightarrow 4H_2 + NaBO_2 \cdot xH_2O + \text{heat}$$

The traditional hydrolysis of sodium borohydride with liquid water has been extensively studied and the reaction and its kinetic limitations are well documented in literature. Although thermodynamically favored, the aqueous hydrolysis does not proceed to completion even in excess water because of the formation of basic byproducts that inhibit the reaction. Therefore an acid catalyst is required to adjust the pH and increase the hydrogen yield. In addition, the hydride must be stored as a caustic solution to prevent premature reaction.

Alternatively, our approach is to expose solid chemical hydrides to steam. We have demonstrated essentially 100% yield without the need for a catalyst and without the need for storage in caustic solution. Clearly, the reaction mechanism and the pathway for steam hydrolysis differ from traditional liquid hydrolysis.

The purpose of this study is to elucidate the mechanism and physical phenomena associated with steam hydrolysis of sodium borohydride an alternative pathway that may significantly improve hydrogen storage by chemical hydrides. In-situ visual observation of the reaction and its phase behavior was accomplished with a glass reactor and borescope camera. Reaction products were analyzed by 11B NMR from both the steam and aqueous hydrolysis reactions for comparison. The reaction rate and conversion appear to rely on the ability of the hydride to absorb water vapor and then dissolve in it, a phenomenon called deliquescence. This process depends on the reaction temperature, pressure and mole fraction of water. Based on this study, deliquescence is necessary for reaction, but only occurs above 30% humidity at 110oC and requires higher humidity as reaction temperature is increased. Analysis of the reaction products indicates that a BH3 containing intermediate found in the aqueous reaction is not present in the steam pathway, indicating a different reaction mechanism. Comparison of liquid and vapor phase hydrolysis provides valuable insight into the reaction mechanisms.

Material Property Study of Thermal Battery Components

Babatunde Ekundayo, Sean Walker, and Irvin Osborne-Lee
Prairie View A&M University Prairie View, TX

ABSTRACT

The aim of this work is to enhance the knowledge base of material properties that characterize thermal battery performance under non-operating and operating conditions; the latter presenting elevated temperature, vibration, shock, and acceleration levels typical of both tactical and strategic applications. Though sophisticated electrical and thermal models to guide some portions of the design of thermal batteries have been developed, some portions needed to integrate mechanical performance with chemical and electrical behavior have been lacking.

In order to enhance and refine modeling and design tools, multiple physical properties of the thermal battery components must be known. Key component physical properties being determined include: the storage modulus, thermal expansion, and coefficient of friction. The components of the thermal battery being examined address many of those needed for a comprehensive study, including electrodes, separators, insulation, and canister materials.

The storage modulus was determined with the use of a dynamic mechanical analyzer (DMA). The single cantilever beam testing process was used for the TA instrument 2980. The method used involved ramping from room temperature to 50 °C, then equilibrating at 50 °C, after which the temperature was ramped by 5 °C per minute to 550 °C. A frequency of 1 hertz and amplitude of 10 μm were used.

The thermal expansion was determined with the use of the Anter Corporation's WorkHorse II dilatometer. The method used involved ramping from room temperature to

50 °C, then equilibrating at 50 °C, after which the temperature was ramped by 5 °C per minute to 550 °C in 25 degree segments. A dwell time of 5 minutes was used between each 25 degree segment. For the electrode samples, thermal expansion test were done under inert conditions by the use of Argon gas.

For the measurement of the coefficient of friction an inclined plane system was used. The coefficient of friction was measured under different conditions. These conditions include with or without added weight, and at room temperature or at an elevated temperature.

Thermal battery packaging and performance requirements are being pushed to the limits. The accumulation of material property data on the components of the thermal batteries will aid in the development of thermal models used to design thermal batteries. This paper identifies the gaps in available data for thermal battery modeling and presents new results for battery component materials from ambient to elevated temperatures.

Density Functional Theory Studies on a Reversible Hydrogen Storage "Li-Mg-B-N-H" System

Nianthrini Balakrishnan, Pabitra Choudhury, Venkat R. Bhethanabotla* and Babu Joseph
Sensors Research Laboratory, Department of Chemical and Biomedical Engineering
University of South Florida, 4202 E. Fowler Ave., ENB 118, Tampa, FL 33620

Many experimental and theoretical studies have been carried out to find hydrogen storage materials with reversible properties, low desorption temperatures[1], accelerated kinetics[1] and optimum storage capacity (at 298 K, the US DOE target for gravimetric hydrogen storage capacity is 6.0 wt %)[2]. Experiments suggest that complex hydrides of *Li, Na, Mg, B, Al* and *N* have such properties[1]. In this study, a density functional theory study with the generalized gradient approximation (GGA) and projected augmented wave method (PAW) is performed to find the hydrogen storage properties for different molar ratios X of MgH_2 in the complex storage material $(LiNH_2)_2$ - $LiBH_4$ - $(MgH_2)_X$. Using *ab-initio* methods, stabilities of the structures are established and thermodynamic properties such as heat of reaction and Gibbs energy are found for each reactant and product in the reaction step for proposed reactions. These are compared with literature values to validate the proposed reactions. The reversibility and storage capacity of these materials will be verified from the desorption behavior observed at different temperatures. These theoretically obtained data can be used to validate the suggested mechanism from experimental observations for the reaction. These efforts are expected to contribute towards identification of suitable hydrogen storage materials.

Motivation

Recent research has gained pace to find a reversible material with efficient hydrogen storage properties. Metal hydrides as hydrogen storage materials have gained importance in the past few years. Storage properties of metal hydrides were found to improve in their combinations. Undesired properties of these hydrides such as ammonia liberation, poor kinetics and irreversibility were overcome by their binary systems[1]. $LiBH_4$-MgH_2 system was found to have 8-10 wt% reversible hydrogen storage at about 350 °C[3]. $LiNH_2$-MgH_2 system operates at 200 °C with 4.5 wt% reversible hydrogen storage[4]. Experiments on the ternary systems of these hydrides also showed promising results and further improved their storage properties. The combined $LiNH_2$-MgH_2-$LiBH_4$ system was found to have lowered desorption temperatures, increased hydrogen purity, accelerated kinetics and partial reversibility[1]. The stoichiometry of this mixture was also a major factor controlling the storage properties.

DFT studies on hydrogen storage materials using projected augmented wave (PAW) potentials pertaining to generalized gradient approximation (GGA) are considered to be the most appropriate for energy calculations due to their better accuracy than Local Density Approximation (LDA)[5]. The thermodynamic properties calculated using plane

* Corresponding author email: venkat@eng.usf.edu

wave DFT and those found experimentally are in good agreement for a number of reversible hydrogen storage systems[6].

Computational details

In this work, we tested the hydrogen storage properties of the $(LiNH_2)_2$ - $LiBH_4$ - $(MgH_2)_X$ reactant mixtures using DFT studies. The reactants and products of the proposed reaction mechanism were checked for stability and then the thermodynamic properties of the compounds were calculated and compared with literature. We used VASP (Vienna Ab Initio Simulation package) to perform the DFT calculations and the PHONON[8] code to perform lattice dynamics for finite temperature crystal stability and thermodynamics.

The most favored structure of each compound was found from literature and its equilibrium volume was calculated by a quadratic fit of energy. These calculations were performed using Monkhorst Pack meshes containing 4 x 4 x 4 *k*- points. The energy cutoff for all the compounds was set to 600 eV. For the volume relaxations the cutoff energy was increased to about 800 eV. The external pressures acting on each of these compounds were reduced by relaxing the volume and ionic positions of the structure. PAW potentials were used for these relaxations. The relaxations were continued till the forces on the atoms were reduced to about 0.1 meV A^{o-1}. Hellmann–Feynman (HF) forces were calculated using a super cell of the relaxed structure to avoid the interaction between the atoms and then by displacing its atomic positions.

The calculated HF forces were then used to find the phonon density of states (DOS) for the structure at finite temperatures, which would be positive for a stable structure. Entropy and internal energy of the structure can be obtained from the phonon DOS, which can be used to calculate the Gibbs energy of the reactions. The reversibility and hydrogen storage capacity of the reaction for different mole ratios can then be calculated. The experimentally observed XRD pattern will be compared with the theoretically observed ones to validate the structure of the newly identified phase ($LiMgBN_2$). To specify the atomic positions of the compounds with fractional occupancies such as $Li_2Mg(NH)_2$ and $LiMgBN_2$,we used the super cell approach.

Results/Discussion

In this study, we calculated the equilibrium volumes for all the compounds, namely, $LiNH_2$, $LiBH_4$, MgH_2, $Mg(NH_2)_2$, LiH, H_2, Li_3BN_2,Mg_3N_2, $LiMgBN_2$, $Li_2Mg(NH)_2$ and $Li_4(NH_2)_3(BH_4)$ involved in the reaction mechanism. We reduced the pressure acting on the compounds to less than -0.1 kB. The stability of about 8 compounds was tested by using DFT to calculate the HF forces, which were then used in lattice dynamics calculations to derive the phonon spectra, which allowed us to establish stability of these structures at finite temperature

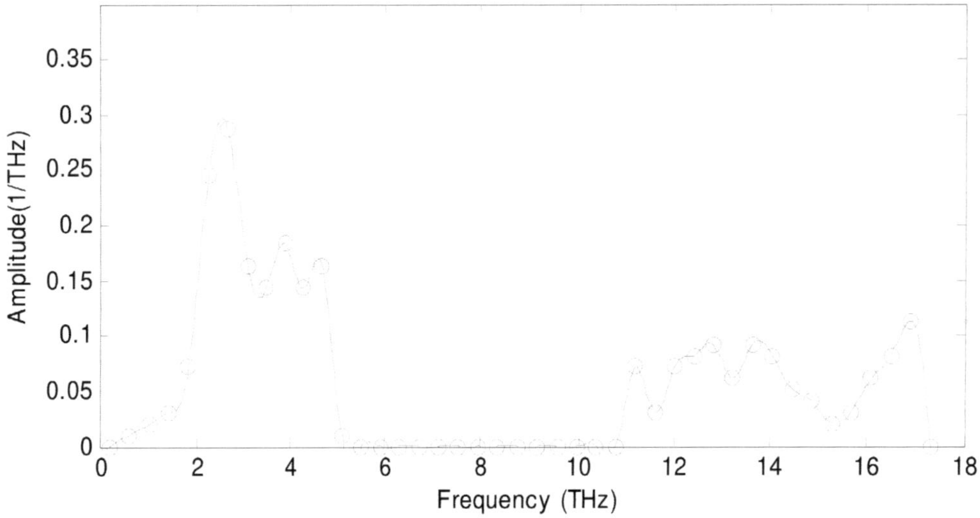

Figure 1: Phonon DOS for Mg_3N_2

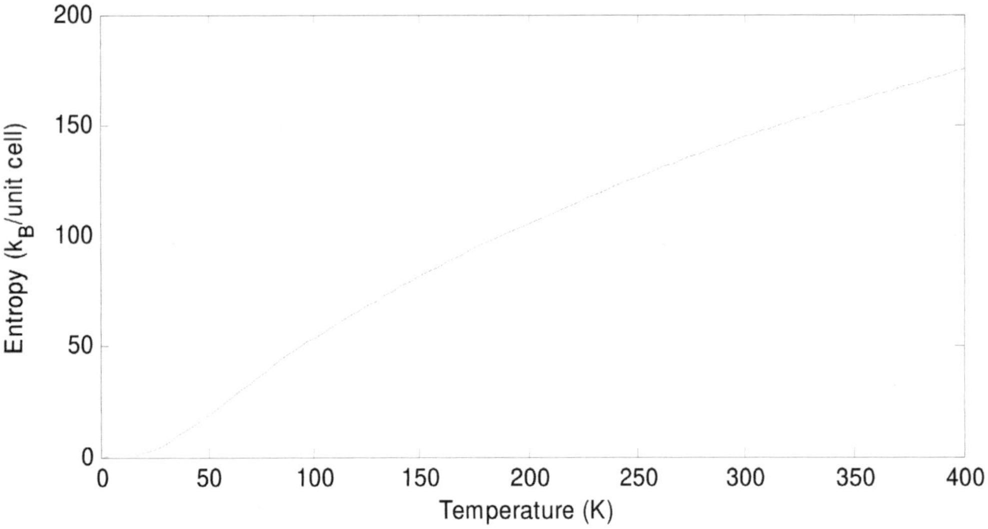

Figure 2: Entropy as a function of temperature for Mg_3N_2

These stable structures will be used to calculate the thermodynamic properties of the reactions. Experimentally proposed multistep hydrogen release reactions for the intermediate quaternary phase $Li_4(NH_2)_3(BH_4)$[1] that is formed will be examined, and the impact of MgH_2 in the storage material for X = 0, 0.25, 0.5, 0.75 and 1.0 will be calculated. The amount of hydrogen desorbed for different mole ratios will also be calculated.

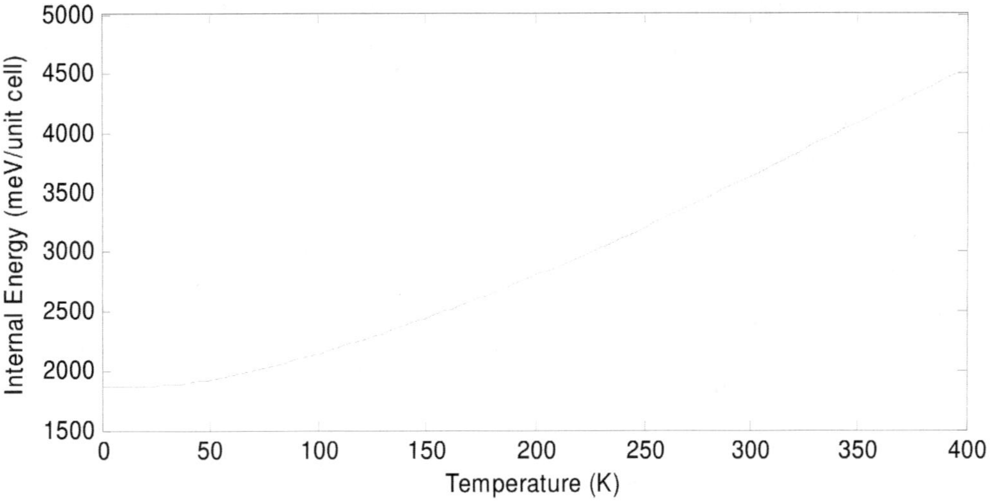

Figure 3: Internal Energy as a function of temperature for Mg_3N_2

Conclusion

The stability of the structures involved in a hydrogen desorption reaction set was evaluated. The thermodynamic properties of these compounds will be calculated and the feasibility of the mechanism will be validated. This DFT study could supplement experiments in testing the metal hydrides for hydrogen storage properties.

References:

[1]Sudik, A.; Yang, J.; Halliday, D.; Wolverton, C., Hydrogen Storage Properties in *$(LiNH_2)_2$ - $LiBH_4$ - $(MgH_2)_X$* Mixtures. *J. Phys. Chem. C* **2008,** 112, 4384.
[2] Satyapal, S.; Petrovic, J.; Read, C.; Thomas, G.; Ordaz, G., The U.S. Department of Energy's National Hydrogen Storage Project: Progress Towards Meeting Hydrogen-Powered Vehicle Requirements. *Catalysis Today* **2006,** 120, 246.
[3]Vajo, J. J.; Skeith, S. L., Reversible Storage of Hydrogen in Destabilized *$LiBH_4$*. *J. Phys. Chem. B* **2005,** 109, 3719.
[4]Luo, W., (*$LiNH_2$–MgH_2*): A Viable Hydrogen Storage System. *J. Alloys Compd.* **2004,** 381, 284.
[5]Barrera, G. D.; Colognesi, D.; Mitchell, P. C. H.; Ramirez-Cuesta, A. J., LDA or GGA? A Combined Experimental Inelastic Neutron Scattering and Ab Initio Lattice Dynamics Study of Alkali Metal Hydrides. *Chem. Phys.* **2005,** 317, 119.
[6]Alapati, S. V.; Johnson, J. k.; Sholl, D. S., Using First Principles Calculation to Identify New Destabilized Metal Hydride Reactions for Reversible Hydrogen Storage. *Phys. Chem. Chem. Phys.* **2007,** 9, 1438.
[7]G.J.Lewis; J.W.A.Sachtler; J.J.Low; D.A.Lesch; S.A.Faheem; P.M.Dosek; L.M.Knight; L.Halloran; C.M.Jensen; Jun Yang; Sudik, A.; J.Seigel, D.; Wolverton, C.; Ozolins, V.; Zhang, S., High Throughput Screening of the Ternary *$LiNH_2$-MgH_2-$LiBH_4$* phase Diagram. *J. Alloys Compd.* **2007,** 446, 355.
[8]Parlinski, K. software PHONON, 2005.

583a Investigation of Esterification of Biomass Pyrolysis Acids Over Sulfonic Acid-Functionalized Mesoporous Silicas

Shaojun Miao and Brent H. Shanks. Department of Chemical and Biological Engineering, Iowa State University, Ames, IA 50010

Esterification of carboxylic acids with alcohols is of great industrial interest due to the enormous practical importance of organic ester products. Additionally, esterification also provides a potential route to stabilize bio-oil produced from the fast pyrolysis of biomass since carboxylic acids are a key constituent in bio-oil that can catalyze reactions such as esterification, acetalization and polymerization, which results in the instability of bio-oil during aging. Compared with homogeneous catalysts such as sulfuric acid, heterogeneous acidic catalysts would be advantageous due to easier separation and recyclability. In the strategy for developing highly active heterogeneous acidic catalyst, organosulfonic acid-functionalized mesoporous silicas are proven to be promising candidates. In this presentation, detailed studies on the kinetics and mechanism of acetic acid esterification over organosulfonic acid-functionalized SBA-15 are discussed, since acetic acid is a prevalent carboxylic acid in bio-oil.

Varying concentrations of sulfonic acid moieties were incorporated into the mesopores of SBA-15 via the one-step co-condensation method. The reaction performance of the functionalized mesoporous silicas was evaluated for the esterification of acetic acid with methanol. In the kinetic studies, the reaction orders for acetic acid and methanol as well as the activation energy were determined. A series of pyridine poisoning experiments were performed to establish the reaction mechanism of esterification over SO3H-SBA-15, i.e., single-site or dual-site mechanism. Since water is a significant component in bio-oil, an investigation of the effect of water on the esterification was performed. SO3H-SBA-15 was found to be more water tolerant than the homogeneous catalyst H2SO4 through a series of experiments with addition of varying levels of water, suggesting that the heterogenous catalyst is a better candidate for pretreating bio-oil.

Oligomerization of Propionaldehyde on Novel Basic Catalysts

Q. V. Amen
Graduate Research Assistant
Chemical, Biological, and Materials Engineering
University of Oklahoma

T. Sooknoi
Faculty of Science
Department of Chemistry
King Mongkut's Institute of Technology Ladkrabang

L.L. Lobban
Chemical, Biological, and Materials Engineering
University of Oklahoma

R.G. Mallinson
Chemical, Biological, and Materials Engineering
University of Oklahoma

Daniel E. Resasco
Chemical, Biological, and Materials Engineering
University of Oklahoma

Abstract

Vapor-phase oligomerization of propanal to 2-methyl-2-pentenal and 2-ethyl-3,5-dimethylcyclo-2-pentenone over basic zeolite catalysts (NaX, KNaX, KX, and CsNaX), as well as more traditional solid basic catalysts (MgO and MgO-Al2O3 hydrotalcites), was studied in an integral flow reactor at atmospheric pressure. Conversion and carbon yield based on catalyst properties (basicity, acidity) are reported. The highest stable activity was over KX, with the highest selectivity to the primary product, 2-methyl-2-pentenal (dehydrated aldol dimer of propanal) and the second highest selectivity to the secondary product, 2-ethyl-3,5-dimethylcyclo-2-pentenone (a non-aromatic cyclic trimer), with essentially no other products. Promotion of Zeolite X with various alkali metals also affected the acid and base site strengths, which had an effect on the conversion and selectivity.

Keywords: Propanal; Aldol condensation; Solid base; Basic Zeolite; Cesium

1. Introduction

With the increased price of petroleum around the world, countries with high consumption or low reserves are interested in producing transportation fuels from other sources. Producing liquid fuels from biomass has the advantage of being local, both decreasing dependence on foreign powers as well as creating local jobs.

One popular process for biomass conversion involves the transesterification of vegetable oils, using methanol, to produce fatty acid methyl esters (FAMEs) with a glycerin (impure glycerol, 1,2,3-propanetriol) byproduct stream. Although a market for purified glycerol does exist, it will be overwhelmed if all the planned biodiesel facilities purify and sell the resultant glycerol. The result would be a precipitous drop in the price of what was intended to be a side revenue stream, resulting in negative economics for equipment installation and energy consumption of glycerol purification. An alternative market must be found for this high-volume stream, preferably with positive economics.

Thermal and catalytic decomposition of glycerol produces acrolein, which may be hydrogenated to propanal with a reactive aldehyde moiety. The well-known aldol condensation reaction, that increases hydrocarbon molecular weight via carbon-carbon bond formation, is catalyzed in solution both by acids and by bases, so it is a reasonable proposition to study the use of solid acids or solid bases to catalyze these reactions. Since solid acids tend to promote cracking, and given the appreciable activity of basic catalysts at low temperatures, solid base catalysts have been studied in this work at low enough temperatures to avoid cracking.

The dimer and trimer species resulting from the propanal condensation reaction, in which each addition of a monomer is accompanied by a dehydration step, are potentially valuable components of the gasoline pool. The dimer could be hydrogenated and deoxygenated, while the (cyclic) trimer is a highly branched molecule, potentially with a high octane number. While the intermediate dehydration steps remove one oxygen after each condensation step, the ring closure step does not eliminate an oxygen atom. This oxygen remains in a carbonyl group, but it should not be overly reactive in the fuel pool: the C-O double bond is conjugated with a C-C double bond, so it should be less reactive than unconjugated carbonyl compounds. In any case, if necessary, the carbonyl and alkene moieties in the trimer can be hydrogenated to a saturated alcohol, or even deoxygenated altogether in a subsequent step.

In this contribution, the activity, selectivity, and stability of a series of basic catalysts with or without associated mild acidity have been compared for the self-condensation of propanal. As will be seen below, the alkali-ion-exchanged zeolite X catalysts generally have both higher activity and higher stability than the classical basic catalysts. The resulting trimer/dimer ratio has been found to be a strong function of conversion, but also the catalyst type.

2. Experimental

2.1. Catalyst preparation

Several alkali-ion-exchanged zeolite catalysts were compared in the propanal condensation reactions. The NaX zeolite was used as purchased (Molecular Sieve UOP Type 13 X from Fluka, product no. 69856). From this were prepared KX and CsNaX by ion exchange.

To prepare the magnesium oxide (MgO), typically 12 grams of magnesium nitrate hexahydrate (99% A.C.S. reagent, Sigma Aldrich) and 6 g of citric acid were dissolved in approximately 60 mL of water in a thin walled mortar. This was then placed in a 500°C oven for 2 h. The resulting solids were ground lightly with a pestle.

The hydrotalcite (HT) was used as purchased (synthetic, from Sigma Aldrich).

2.2. Fresh Catalyst Characterization

Temperature programmed desorption (TPD) of adsorbed CO_2 was used to explore the basic nature of the catalysts, while TPD of adsorbed isopropyl amine (IPA) was used to explore the acidic nature of the catalysts. A mass spectrometer was used as the detector for TPD experiments.

2.3. Measurement of Catalytic Activity

The gas phase down-flow reactor was built of 3/8" stainless steel tubing with an internal temperature controlling thermocouple. Quartz wool was placed at the thermocouple to form a support for the catalyst bed. After adding the catalyst (typically 300 mg), more quartz wool was placed on top of the bed.

The liquid feed was propanal (reagent grade, >97%, Sigma Aldrich). To prevent oxidation, the head space was swept with He whenever the bottle was opened. To study the effect of feed oxidation, propanoic acid (99.5+% A.C.S. from Sigma Aldrich) was used in some runs.

Liquid feed was delivered to the system at 0.5 mL/hr by a syringe pump with a needle pointing down through a septum into the reactor. The reactor effluent line was heated to 220°C and fed through a 6-port sampling valve. The vent line from the 6-port valve could be submerged in methanol to collect products for GCMS analysis on a Shimadzu GCMS-QP2010S. A sample loop was used to capture a snapshot of the effluent gas, which was fed to an Agilent 6890 for GC analysis via FID through a 30m HP5 column or a 30m Innowax column. Due to the presence of heteroatoms in the feed and product molecules, a response factor for each compound was used to determine the carbon fraction rather than using the raw FID signal.

Zeolites were pretreated by heating from ambient to 450°C at 4°C/min in a flowing air and helium mixture (air 7.0 mL/min, helium 45.6 mL/min) and held at 450°C for 1 hour. The reactor was then cooled to 300°C, the air flow was eliminated, the helium flow was set to 39.1 mL/min, and the reaction was initiated by starting the syringe pump.

MgO and HT were pretreated by heating from ambient to 500°C at 10°C/min in a flowing air and helium mixture (air 7.0 mL/min, helium 45.6 mL/min) and held at 500°C for 2 hours. The reactor was then cooled to 300°C, the air flow was eliminated, the helium flow was set to 39.1 mL/min, and the reaction was initiated by starting the syringe pump.

The chemical reaction investigated was aldol self-condensation of propanal. While aldol condensation is well-known, and some researchers have used propanal as a reactant in cross-condensation, few researchers have studied propanal self-condensation in any detail. Drawing an analogy from the reaction catalyzed by a homogeneous base, one would expect the first step our system of solid bases to be proton abstraction. The resulting carbanion donates its lone pair to the carbonyl carbon of an adjacent propanal molecule, pushing the □ electrons onto the oxygen. The negatively charged oxygen atom is then reprotonated, and the subsequent facile dehydration results in the □ □ unsaturated carbonyl compound (here called the "dehydrated dimer").

The next reaction step to the linear trimer follows the same mechanism. For acetone self-condensation, two linear trimer species are possible, but it does not matter whether propanal or the dehydrated dimer is converted to the enolate-like species; both options produce the same linear trimer, 2,4-dimethylhepta-2,4-dienal. This molecule was not identified by GCMS, nor were there any unidentified peaks in the GC traces which could be attributed to this species. However, GCMS did identify a 9-carbon cyclic species (2,3,4,5-tetramethyl-2-cyclopentenone) similar to one which may be formed via Nazarov cyclization (2-ethyl-3,5-dimethyl-2-cyclopentenone), which is catalyzed exclusively by acids. It is believed that the actual species is indistinguishable by GCMS from the tetramethyl isomer, due to the improbability of producing an ethyl fragment from the ethyl-dimethyl isomer.

In the alkali-exchanged zeolites, the alkali metal cations act as adjacent acid sites, so the resulting enolate oxygen (with a formal -1 charge) may be stabilized by electron donation to the cation. Alternatively, when acid-base pairs are present, the oxygen of the carbonyl could adsorb onto alkali metal cations, followed by proton abstraction by the adjacent base site. As mentioned above, it does not matter which molecule acts as the enolate-like species; the same linear trimer is formed.

While the linear trimer is not observed in the GC or GCMS analysis, a cyclic trimer is observed. If the Nazarov electrocyclic ring closure is the correct mechanism, then the Lewis acidity of the alkali cations must catalyze this step of the reaction scheme. Solids with only basic character would not be expected to produce any of this trimer.

2.4 *Spent Catalyst Characterization*

Temperature programmed oxidation (TPO) experiments were conducted on the spent catalysts to determine the nature of the carbonaceous species remaining after reaction. A small amount (about 40 mg) of spent catalyst was loaded in a ¼" quartz tube. A dilute mixture of oxygen in helium was flowed over the catalyst while the temperature was increased at 10°C/min up to 900°C. The effluent gases flowed through a methanator to convert the CO2 to methane, which was then fed to an FID detector. After the TPO run, a sample loop filled with CO_2 was cycled to produce reference peaks.

3. Key results

Relative activities of all catalysts studied appear in Figure 1. Basic zeolites are generally superior to the traditional solid bases in terms of activity per gram and longevity, except for the case of NaX which deactivates rapidly. Other authors discussing solid bases only provide data up to one hour Time On Stream (TOS), or do not report what the TOS is.

In addition to the longevity, another remarkable feature of the zeolites is their resistance to poisoning by the oxidized feed, apart from NaX, whose conversion is reduced from 65% with clean feed at 0.75 h TOS to only 19% conversion (a 70% drop) at 1 h TOS with the oxidized feed. In contrast, the KX catalyst exhibits similar initial activity (41% at ~1 h TOS) in both cases, although KX deactivates more quickly than with clean feed. In contrast to both of the other alkaline zeolites, CsNaX exhibits both augmented activity and better stability with PA in the feed (Figure 2). While not having as high of an activity compared to the zeolites, MgO shows measurable activity for several hours TOS when clean feed is used. However, MgO is completely deactivated within an hour when PA is present, while most of the zeolites continue with appreciable activity beyond five hours, even under the least favorable conditions (with PA for NaX and KX, without PA for CsNaX.

Figure 3 shows the ratio of trimer yield to dimer yield as a function of conversion for all catalysts. The CsNaX clearly shows a higher tendency to produce trimer than the other catalysts. The MgO, HT, and KX all fall roughly along the same line, while the NaX data fall between the other catalysts.

4. Acknowledgements

Oklahoma Secretary of Energy

Oklahoma Bioenergy Center

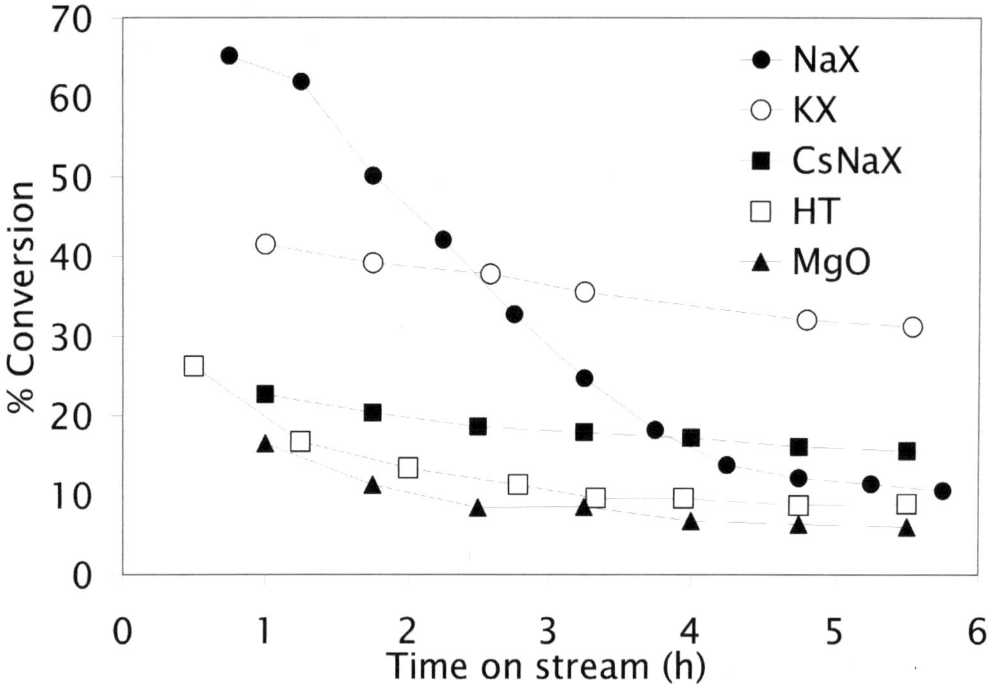

Figure 1. Conversion over base catalysts as a function of time on stream. Reaction conditions: 300°C, 1 atm, 0.5 mL/hr clean propanal; 0.3 g catalysts; W/F = 0.745 hr.

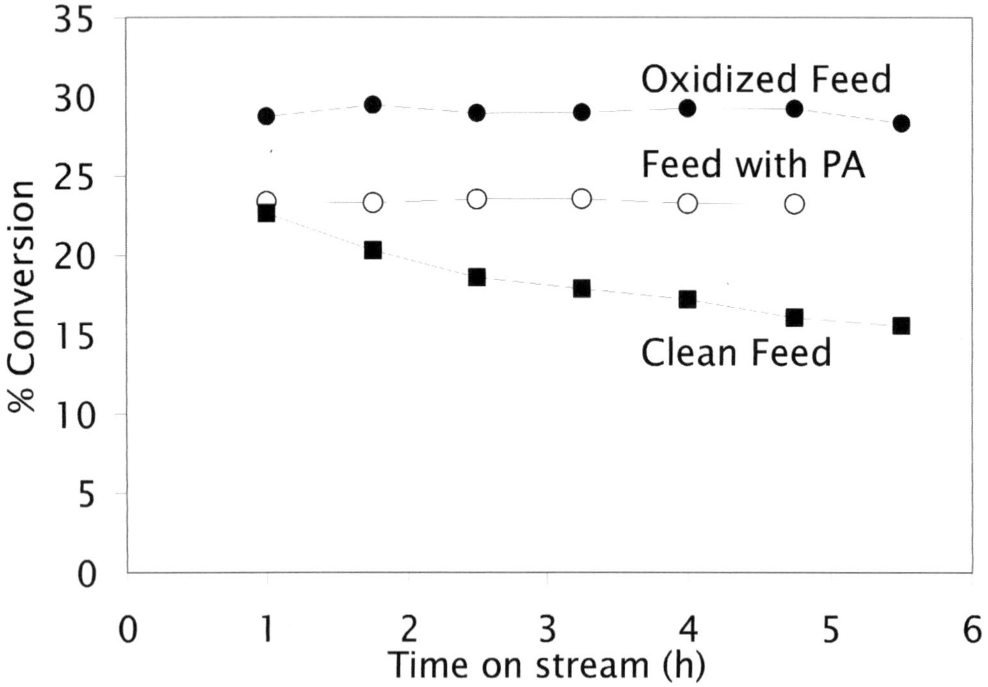

Figure 2. Conversion over CsNaX as a function of time on stream. Reaction conditions: 300°C, 1 atm, 0.5 mL/hr propanal; 0.3 g catalysts; W/F = 0.745 hr

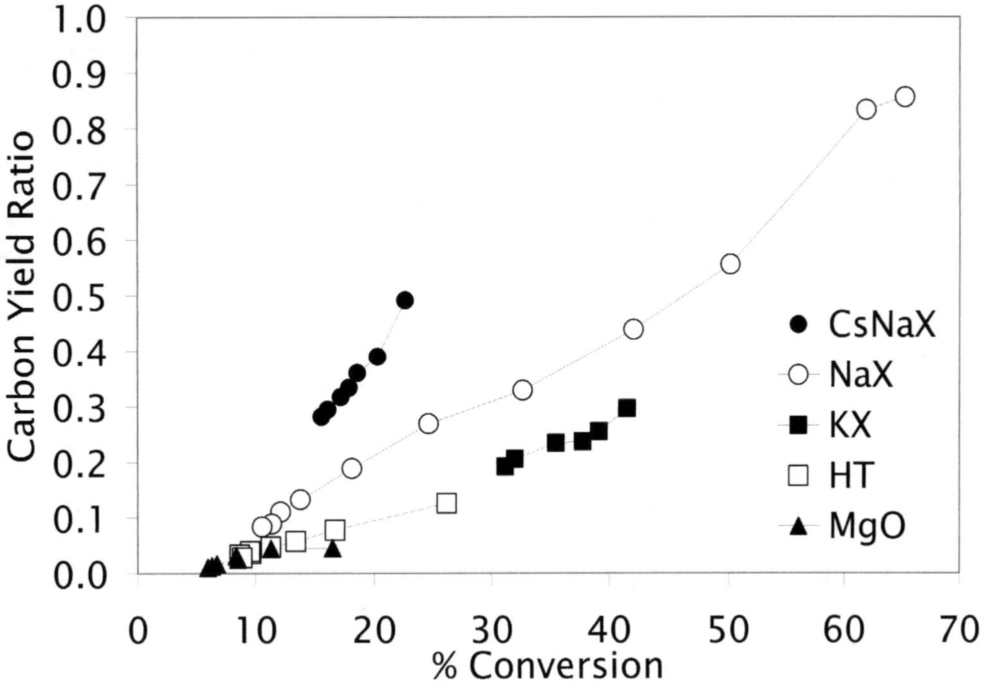

Figure 3. Ratio of trimer yield to dimer yield over base catalysts as a function of conversion. Reaction conditions: 300°C, 1 atm, 0.5 mL/hr clean propanal; 0.3 g catalysts; W/F = 0.745 hr.

583c Comparative Analysis of Biodiesel Production from Waste Cooking Oil Using Ferric Sulfate and Sulfuric Acid

Prafulla Patil, New Mexico State University, P.O. Box 30001, MSC 3805, Las Cruces, NM 88003-8001 and Shuguang Deng, Chemical Engineering Department, New Mexico State University, New Mexico State University, P.O. Box 30001, MSC 3805, Las Cruces, NM 88003.

In this comparative study, biodiesel production from waste cooking using ferric sulfate and sulfuric acid catalyst was carried out. The two-step transesterification process was used to get rid of high free fatty acid (FFA) content in waste cooking oil. This process gives yield of about 85-95% using sulfuric acid catalyst and 80-98% using a ferric sulfate catalyst. Ferric sulfate acted as a heterogeneous acid catalyst and showed good activity to catalyze the methanolysis of FFA in the waste cooking oil. This catalyst is environmental friendly, easy to separate and more efficient than sulfuric acid catalyst. The most important variables affecting the methyl ester yield during the transesterification reaction are the molar ratio of alcohol to oil, the catalyst amount and the reaction temperature. The analysis of different oil properties, fuel properties and process parameter optimization of waste cooking oil were studied in detail. The fuel properties of biodiesel produced were compared with ASTM standards for biodiesel and regular petroleum diesel.

583d Impact of Oxygen Incorporation In Molecular Structure on Pyrolytic Soot Formation and Oxidizability

Steven P. Crossley, Maria Pilar Ruiz, Liang Zhang, and Daniel E. Resasco. Cbme, University of Oklahoma, 100 East Boyd, Room T-335, Norman, OK 73019

Recent advances towards the use of renewable sources of energy warrant further investigation of the effect of oxygenated hydrocarbons on soot formation and reactivity. The recently developed tool, the micropyrolysis index (MPI)[1] was used as a basis for these studies. Carbon was deposited on same spot silicon nitride grids by placing the grid in a MPI flow reactor at a 45° angle from the direction of flow and injecting 20 μ L of specific fuels in a stream of He across the grids at a temperature of 1000°C. Pyrolized particles deposited directly on the grids were analyzed via TEM and Raman. Partial oxidation of the grids were then conducted in a temperature programmed oxidation (TPO) reactor, with resulting analysis of the same particles via same spot TEM and Raman. Compounds investigated include toluene and benzaldehyde, in order to study the effect of oxygen on the soot formation via pyrolysis and reactivity. Reactivity was studied with and without the presence of an oxidation catalyst (CeO_2), allowing one to study the influence of both starting molecular structure and catalytic influence.

Similar studies were conducted by replacing the TEM grid with a bed of α-Al_2O_3 (same method used for MPI measurements), and analyzing the deposited carbon via TPO and SEM of the particles. Through this method, several compounds were tested, including benzene, benzaldehyde, dodecane, and methyl-dodecanoate. TPO measurements indicate a drop in the amount of soot produced from oxygenates when compared with their corresponding hydrocarbons. Also, the appearance of a high temperature peak is observed to be much more significant for toluene than for benzaldehyde, indicating larger, more graphitic particles. By combining the two studies, findings indicate that the effect of oxygen on soot formation via pyrolysis is to lower the total amount of soot, and thus average particle size through radical scavenging as the unstable oxygen species produced via pyrolysis act as radical scavengers, reducing the average particle size and total amount of soot produced.

1. A Novel Micropyrolyis Index (MPI) to Estimate Sooting Tendency of Fuels, Steven P. Crossley, Walter E. Alvarez, Daniel E. Resasco. Accepted: Energy and Fuels.

583e Conversion of Acetone to Hydrocarbon on Titania Based Catalysts

Prashant Reuben Daggolu, Dave C. Swalm School of Chemical Engineering, Mississippi State University, Swalm 258, Mississippi State University, Mississippi State, MS 39762 and Mark G. White, Chemical Engineering, Mississippi State University, Mississippi State, MS 39759.

The conversion of biomass-derived syngas into a gasoline blend can be achieved by an alcohol route through the familiar TIGAS technology in which one of the intermediates is methanol. It is now known that when syngas is converted to intermediates that are higher alcohols (longer carbon chain than Methanol), dehydration of these species give higher yields of gasoline that what was observed when methanol is the intermediate. One catalyst that can convert syngas to higher alcohols is rhodium, but the rhodium catalysts also form ketones in significant yields among other oxygenates, apart from higher alcohols. Any commercially-viable technology involving this pathway must provide for the complete conversion of oxygenates to hydrocarbons. Hence, we studied the aldol condensation of acetone to mesitylene in high pressure conditions in hydrogen-rich atmospheres. These conditions were chosen as way to mimic the conditions inside commercial reactors during syngas conversion, where hydrogen is readily available.

Acetone is known to undergo self condensation to form mainly mesitylene with isophorone, xylenols and mesitylene oxide as main by-products on the same catalyst. Mesitylene is a hydrocarbon which can be left in the fuel stream, if modest yields of aromatics can be tolerated. Acetone is also known to form more varied aromatics on zeolites. Studies are performed on combining titanium with zeolites, such as HZSM-5, to form gasoline blends containing benzene, xylene, toluene and some aliphatic short chain hydrocarbons, from acetone. The yields obtained over physical mixtures of titania and HZSM-5, impregnated catalysts and separated beds will be compared. This combination will be optimized for direct conversion of acetone to gasoline range hydrocarbons.

Studies will be shown on these reactions at different conditions of pressure (0 - 1500 PSIg), temperature (175 – 475oC) and flow rates(0.1 -1.5 cm3/min) with pure, poly-crystalline titania catalyst. The best conditions for acetone conversion to mesitylene will be elucidated. Also, some insight shall be given into the mechanism of the reactions.

DEOXYGENATION OF MODEL BIOFUEL COMPOUNDS OVER CsNaX ZEOLITE CATALYST

Tanate Danuthai[1], Maria A. Peralta[2], Tawan Sooknoi[3], Somchai Osuwan[1], Lance L. Lobban[2], Richard G. Mallinson[2], and Daniel E. Resasco[2]

[1] The Petroleum and Petrochemical College, Chulalongkorn University, Bangkok, Thailand, 10330
[2] School of Chemical, Biological, and Materials Engineering, The University of Oklahoma, Norman, OK, 73019
[3] Department of Chemistry, Faculty of Science, King Mongkut Institute of Technology Ladkrabang, Bangkok, Thailand, 10520

ABSTRACT

The deoxygenation of methyl octanoate and benzaldehyde over CsNaX zeolite catalyst has been investigated as a model for reactions for production of hydrocarbon fuels from bio-oxygenates. The CsNaX zeolite used in the study was prepared by ion exchange of NaX with a $CsNO_3$/CsOH solution to a 50% exchange level. The decarbonylation activity, selectivity, and stability of the CsNaX catalyst were enhanced when methanol was co-fed for the reaction of methyl octanoate. TPD studies suggest that methyl octanoate first decomposes to an octanoate-like species. The decomposition of such species leads to the formation of heptenes and hexenes as major products. Octenes and hydrogenated products are formed in lesser amounts via hydrogenation / hydrogenolysis using hydrogen produced on the surface from methanol decomposition, but not from gas phase H_2.

INTRODUCTION

The interest in renewable fuels as replacements for fossil fuels has rapidly increased over the past few years. Among the various fuels-from-biomass investigated, fatty acid methyl-ester biofuels (FAME) obtained by trans-esterification of triglycerides from natural oils and fats with methanol and production of bio-oil by flash pyrolysis of lignocellulosic biomass have received considerable attention [1-4]. Due to the oxidation and thermal instability of these biofuels, there is continuing concern about their fungibility. Reduction of the reactive oxygen content in the fuel would improve the stability of the fuel and therefore its utilization potential. Various processes including hydrogenolysis [5,6], decarbonylation [7,8], and decarboxylation [9,10] have been proposed to transform the biofuels into a fungible hydrocarbon base fuel.

Typically, decarbonylation takes place over supported noble metal catalysts. For example, Pd/C has been found to be an effective catalyst for decarbonylation / decarboxylation [9]. However, CO produced from the reaction may competitively adsorb on the metal surface, leading to a loss in catalytic activity as conversion increases. Hence, total and hydrogen partial pressures are generally high, in order to keep the surface clean and facilitate the oxygenate-metal surface interaction, and reduce catalyst deactivation [11]

An alternative family of catalysts that may overcome some of these limitations is that of solid bases because the intermolecular interactions and adsorption site competition may greatly affect the reactivity of a catalyst. Among possible solid base catalysts, low-silica zeolites containing highly polarizable cations, such as cesium, may be good candidates.

These catalysts have been found to exhibit relatively high activity towards hydrogenation and hydrogenolysis of acrylonitrile and propionitrile [12]. Recent studies [13] suggest that, in addition to the need for high base strength, an active catalyst requires the co-existence of acid and basic sites for reactions such as hydrogenation and hydrogenolysis.

The purpose of this work is to investigate deoxygenation reactions on Cs- and NaX zeolite catalysts, using methyl octanoate as a model biofuel compound. The effects of varying reaction conditions such as temperature, hydrogen partial pressure, feed concentration, presence of competing molecules, as well as catalyst formulations were investigated in a flow reactor. In addition, temperature programmed techniques were employed to elucidate possible reaction pathways.

EXPERIMENTAL

Catalyst Preparation and Characterization

Molecular sieve (UOP type 13X, NaX) catalyst was commercially obtained from Fluka. The as-received materials were calcined at 450 °C. The Cs-containing zeolite (CsNaX) was prepared by ion exchange of Molecular sieve 13X with 0.1 M $CsNO_3$/CsOH (4/1 v/v) at 80 °C for 24 h. The solid material was filtered and left to dry at 80 °C overnight. The sample was then calcined at 450 °C for 2 h in a flow of dry air. Surface area of the CsNaX and NaX catalysts was determined by Nitrogen adsorption at 77 K using a Micromatics ASAP 2000 apparatus. X-ray diffraction (XRD, Bruker AXS D8Discover) was employed to confirm the catalyst structure.

Temperature Programmed Desorption (TPD) of Methyl Octanoate

The evolution of pre-adsorbed methyl octanoate over CsNaX catalysts was followed by temperature programmed desorption in an apparatus equipped with a mass spectrometer (MS). In each run, 50 mg of the sample was initially pretreated in a flow of 2% O_2/He for 2 h at 450 °C. Then, the sample was cooled in He flow to 150 °C. 100 µl of methyl octanoate was injected to the sample, and a He flow was introduced for 3 hours at 150 °C in order to remove the excess methyl octanoate. The sample was then heated up to 900 °C at a rate of 10 °C/min. Masses (*m/z*) of 2, 28, 56, and 74 were monitored to determine the evolution of hydrogen, carbon monoxide, hydrocarbons, and methyl octanoate, respectively.

Temperature Programmed Desorption (TPD) and Temperature Programmed Reaction (TPRx) of methanol

Methanol was chosen as a probe for the zeolite surface and as a co-reactant to help reduce the rate of catalyst deactivation during the conversion of methyl octanoate. Therefore, the desorption and decomposition of methanol over CsNaX and NaX catalysts was investigated by TPD and TPRx techniques in the same apparatus described in the previous section. In each run, 50 mg of the sample was initially pretreated in a flow of 2% O_2/He for 2 h at 450 °C. Then, the sample was cooled in He flow to room temperature. In the desorption experiment, 10 µl of methanol was repeatedly injected over the sample at room temperature until reaching saturation, as indicated by a constant pulse size MS signal. After the removal of the excess methanol by flowing He for 3 h at room temperature, the sample was then heated to 900 °C at a rate of 10 °C/min. For the TPRx experiment, methanol was continuously fed through the sample bed at a flow rate of 1 ml/hr. The sample was linearly heated from room temperature to 700 °C at a heating rate of 10 °C/min.

Masses (*m/z*) of 2, 28, 31, and 45 were monitored to determine the evolution of hydrogen, carbon monoxide, methanol, and dimethylether, respectively.

Catalytic activity measurements

The reaction was carried out at atmospheric pressure, in a fixed bed flow reactor made with 1/4" quartz tube. Prior to the reaction, the catalyst samples were pretreated *in situ* under flow of 2% O_2/He at 450 °C for 2 h. Then, the catalyst bed was cooled to the reaction temperature (425 °C) under flow of He (25 ml/min). The 10 %wt methyl octanonate in methanol solvent was introduced into the reactor using a syringe pump via a heated vaporization port. Alternatively, in some runs, nonane was used as a solvent instead of methanol. Unreacted feed and products were quantified using an online GC-FID equipped with a capillary HP-5 column, following a temperature program to optimize product separation.

Temperature Programmed Oxidation (TPO) of coke deposits after reaction

Temperature programmed oxidation (TPO) was employed to investigate and quantify the nature and amount of coke deposited on the spent catalysts. 30 mg of the sample was packed in the 1/4" quartz tube reactor. The temperature was ramped to 900 °C with a heating rate of 10 °C/min and the TPO profiles were recorded under flow of 2% O_2/He. The CO_2 produced by the oxidation of the coke deposits was converted to methane over a methanation catalyst (15% Ni/Al_2O_3) in the presence of hydrogen at 400 °C. The corresponding methane was then analyzed online by an FID detector. The amount of oxidized coke was calibrated using 100 μl pulses of pure CO_2 injected to the same system.

RESULTS AND DISCUSSION

Catalytic Activity of the Deoxygenation of methyl octanoate

Effect of Solvents

The reaction of 10%wt methyl octanoate in different solvents (nonane and methanol) was carried out to investigate the effect of solvent. It can be seen from Figure 1 that, the reaction using nonane, a non-polar solvent, gives a high activity at the beginning of the reaction. However, rapid deactivation of the catalyst is observed shortly after an hour on stream. This is presumably because the adsorption of methyl octanoate on the highly basic sites in zeolite is much stronger than that of the hydrocarbon solvent. This results in saturation of the feed in the zeolite micropores and the adsorbed methyl octanoate may well react with itself (self-condensation) forming high molecular weight oxygenates. Such compounds cannot readily diffuse out of the zeolite pores and may well inhibit further feed from adsorption and reaction. Hence, a rapid deactivation was observed after a short period of time on stream. Consistent with this view, the major product from the reaction is mainly 8-pentadecanone, a coupling product of methyl octanoate. Deoxygenated products, namely octene, heptene and hexene, are also observed but with lower selectivity (Figure 2a)

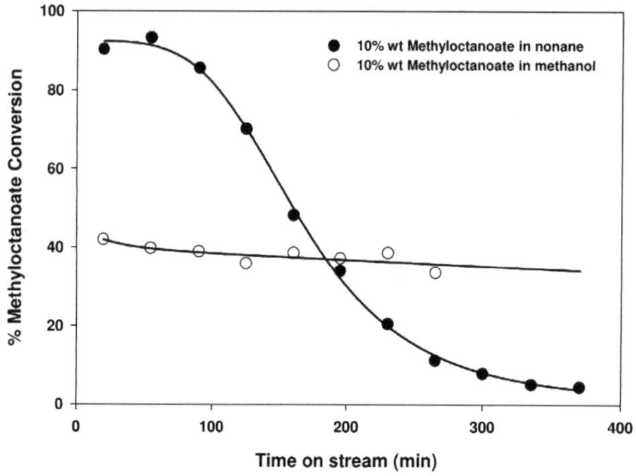

Figure 1 Conversion of 10 %wt methyl octanoate over CsNaX using nonane and methanol as solvent as a function of time on stream. Reaction Conditions: 425 °C, 1 atm, W/F = 198 g*h/mol, 25 ml/min of He.

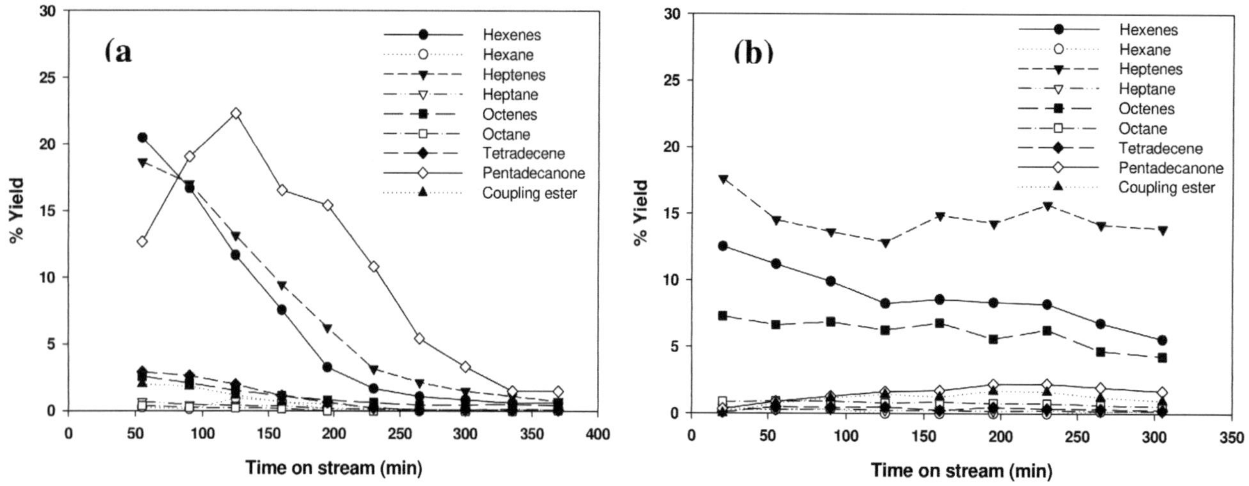

Figure 2. Product distributions from the reaction of 10% wt methyl octanoate using (a) nonane and (b) methanol as solvent as a function of time on stream. Reaction Conditions: 425 °C, 1 atm, W/F = 198 g*h/mol, 25 ml/min of He.

However, when methanol was used as a solvent, a reaction with relatively lower activity but high stability was observed (Figure 1). The explanation is that methanol is highly polar and can competitively interact with the zeolite basic sites. This leads to the decomposition of methanol to give formaldehyde, carbon monoxide and hydrogen. These products can remain on the surface and inhibit the co-adsorption and self-condensation of methyl octanoate. Accordingly, such "virtual" pressure of hydrogen and other methanol products would prevent the formation of higher molecular weight oxygenates resulting in better catalytic stability. The results of decomposition of methanol over CsNaX by TP reaction are shown in Figure 3. It is interesting that a high H_2/CO is pronounced only at 300-400 °C. This indicates that, at this temperature, methanol would primarily decompose to formaldehyde that can readily remain on the surface. As the temperature is increased, the adsorbed formaldehyde can decompose to carbon monoxide and hydrogen as shown by a constant H_2/CO at higher temperature. This adsorbed formaldehyde is likely to be responsible for the reduced activity, but increased stability of the catalyst in the presence of methanol.

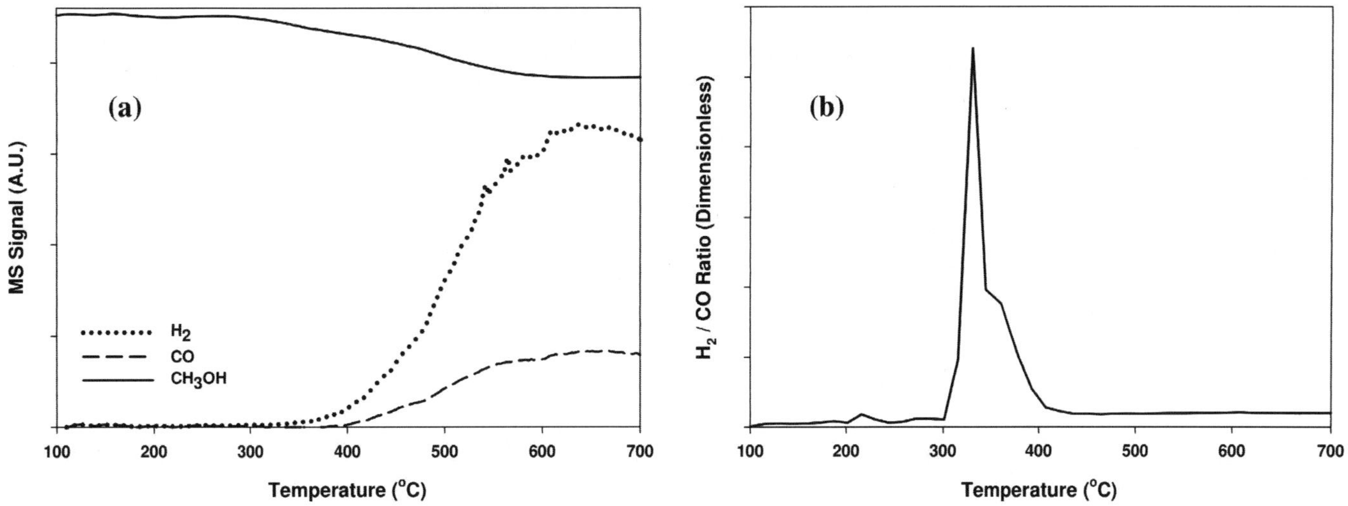

Figure 3. (a) Temperature Programmed Reaction of continuous flow methanol over CsNaX (b) H_2/CO signal of continuous flow methanol over CsNaX.

It is also suggested that the "virtual" pressure of hydrogen can further facilitate the decarbonylation of the ester. Hence, various deoxygenated products are obtained from the reaction using methanol as a solvent (Figure 2b). It is worth noting that the decarbonylation of methyl octanoate over CsNaX requires no additional hydrogen despite the fact that decarbonylated and hydrogenated products are largely obtained. This is because methanol does not only act as reaction media, but it is also serving as a hydrogen source for necessary reactions.

Proposed Reaction Pathways

Although it is certain that the reaction of methanol indeed takes place, the methanol conversion is relatively small (~5%) and product selectivities based on methanol are not considered. Only the carbon yield based on methyl octanoate has been taken into account. It can be seen from Table 1 that heptenes and hexenes are mainly produced from catalytic decarbonylation of methyl octanoate over CsNaX. Octene and larger coupling products, together with small amounts of alkanes and oxygenates, are also observed.

Table 1 Product distribution from decarbonylation of different feeds over CsNaX using methanol as solvent Reaction Conditions: 425 °C, 1 atm, W/F = 396 g*h/mol, 25 ml/min of He.

	10 %wt Methyl octanoate in Methanol
Conversion	62.3
Yield (%)	
Hexenes	19.3
Haxane	0.3
Heptenes	26.1
Heptane	0.8
Octenes	6.7
Octane	1.0
Octanal	0.4
Methyl octanonate	37.7
Oxygenates	3.4
Octanonic acid	2.2
Tetradecene	0.6
Pentadecanone	0.9
Coupling ester	0.8

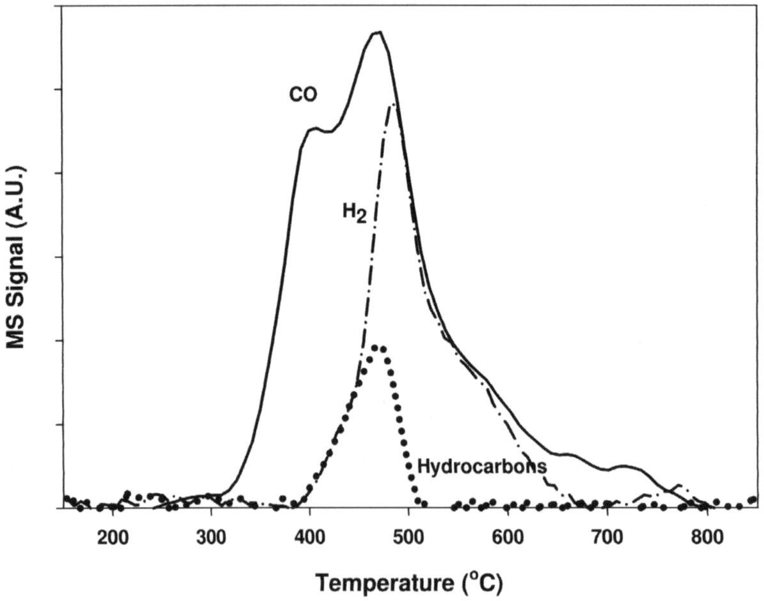

Figure 4. TPD profile of adsorbed methyl octanoate on CsNaX zeolite catalyst

From Figure 4, TPD of methyl octanoate revealed no evolution of the methyl octanoate. Instead, formaldehyde, hydrogen, carbon monoxide and hydrocarbons are evolved above ~300 °C. This suggests that methyl octanoate is strongly adsorbed on the basic sites arising from the negative framework charge countered by the large cesium cation. Due to the highly polarizable nature of methyl octanoate, the interaction between carbonyl ester with the basic framework oxygen would readily weaken the carboxylic C-O bond, leading to the decomposition of the ester into formaldehyde and "aldehyde-like species" at high temperature (>350 °C). The formaldehyde formed would be partly evolved and some would rapidly decompose into carbon monoxide and hydrogen, in a manner similar to methanol

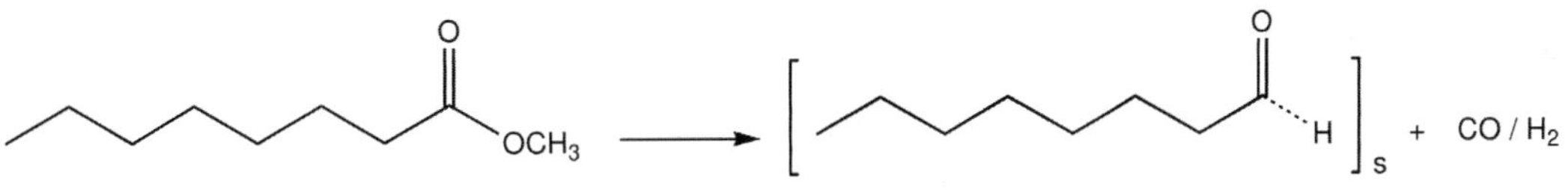

It is very interesting that no hydrogen is evolved until ~500 °C, while hydrocarbons are readily produced. The fact that formaldehyde and carbon monoxide are evolved without hydrogen, leads to the speculation that surface hydrogen would be readily consumed for hydrocarbon formation.

It is proposed that decomposition of such "aldehyde-like species" leads to the formation of two major hydrocarbon products, heptene and hexene. A β-hydrogen elimination [14] and subsequent decomposition to form an olefin, carbon monoxide and hydrogen would give *heptene* as major product of the reaction. This reaction may well be referred to as a reversible reaction of the typical hydroformylation [15]. The decomposition of the cyclic-like intermediate can lead to the formation of more electrophillic acetaldehyde (enol-form) and hexane. The acetaldehyde may be evolved as a product or undergo further aldol condensation/alkylation to form C3-C5 oxygenates, as observed in small amounts by GC-MS

Hydrogen is not essentially required for the formation of heptene and hexene. However, it is largely consumed by the formation of octene and alkanes (Table 1). It is quite plausible to suggest that the formation of octene is derived from additional hydrogenation activity of the cesium exchanged zeolite. As discussed earlier, the basic sites, due to the highly polarizable cesium cation, can readily decompose methanol/formaldehyde, creating a "virtual" hydrogen pressure on the surface. Hence, the adsorbed "aldehyde-like species" would be hydrogenated forming primarily octanol, which may then rapidly undergo dehydration to form octene.

The small amounts of octane, heptane and hexane are further evidence that suggest the hydrogenation activity of CsNaX (Table 1). However, the hydrogenation of octene, heptene and hexene to form octane, heptane and hexane occur to a much lesser extent, as compared to that of the aldehyde-like species to form octanol. This is because olefins are less adsorptive on the basic sites, as compared to the more electrophillic aldehyde-like species. These hydrogenated products can be promoted in the reaction using methanol as a solvent, as discussed earlier.

CONCLUSIONS

It can be concluded that the decarbonylation activity of methyl octanoate can occur at high rate and for a long time on stream over CsNaX catalyst when co-feeding methanol.

Methyl octanoate strongly adsorbs on CsNaX, basic sites and cannot be desorbed unless decomposed. When a less adsorbable solvent, nonane, is used, CsNaX rapidly deactivates. When methanol is present, the catalyst stability is greatly enhanced due to the presence of decomposed fragments of methanol on the surface. The TPD results suggest that the decarbonylation of methyl octanoate proceeds via primary decomposition at the methoxyl group, presumably producing an octanoate-like species as intermediate. The direct decomposition of this species gives heptenes and hexenes as main products. Octenes and other hydrogenated products are formed by hydrogenation / hydrogenolysis, in which the surface hydrogen produced from the methanol decomposition plays an important role.

ACKNOWLEDGEMENTS

This work was financially supported by the Oklahoma Secretary of Energy and by the Oklahoma Bioenergy Center. One of the authors (T.S) thanks the Fulbright Thai visiting program for support, and (T.D.) thanks the Thailand Research Fund for a scholarship.

REFERENCES

[1] D. E. Lopez, K. Suwannakarn, D. A. Bruce, J. G. Goodwin Jr., *J. Catal.* 247 (2007), 43.

[2] N. Martyanov, A. Sayari, *Appl. Catal. A* 339 (2008) 45.

[3] D. A. G. Aranda, R. T. P. Santos, N. C. O. Tanpanes, A. L. D. Ramos, O. A. C. Antunes, *Catal. Lett.* 122 (2008) 20.

[4] M. D. Serio, R. Tesser, L. Pengmei, E. Santacesaria, *Energy and Fuels* 22 (2008) 207.

[5] W. F. Maier, P. Grubmiller, I. Thies, P. M. Stein, M. A. McKervey, P. R. Schleyer, *Angew. Chem. Int. Ed.* 18 (2003) 939.

[6] D. S. Brands, G. U-A-Sai, E. K. Poels, A. Bliek, *J. Catal.* 186 (1999) 169.

[7] T. A. Foglia, P. A. Barr, *J. Am. Oil. Chem. Soc.*, 53 (1976), 737.

[8] I. Kubickova, M. Snare, K. Eranen, P. Maki-Arvela, D. Y. Murzin, *Catal. Today*, 106 (2005) 197.

[9] P. Maki-Arvela, I. Kubickova, M. Snare, K. Eranen, D. Y. Murzin, *Energy Fuels*, 21 (2007) 30.

[10] M. Snare, I Kubickova, P. Maki-Arvela, K. Eranen, J. Warna, D. Y. Murzin, *Chem. Eng. J.* 134 (2007) 29.

[11] M. Snare, I. Kubickova, P. Maki-Arvela, D. Chichova, K. Eranen, D. Y. Murzin, *Fuel* 87 (2008) 933.

[12] T. Sooknoi, J. Dwyer, *J. Mol. Catal. A: Chem.* 211 (2004) 155.

[13] P. Belirame, P. Fumagallip, G. Zuretti, *Ind. Eng. Chem. Res.* 32 (1993) 26.

[14] H. N. Wright Jr., H. J. Hagemeyer Jr., US Patent 3714236 (1973).

[15] N. Takahashi, H. Matsuo, M. Kobayashi, *J. Chem. Soc., Faraday Trans. 1*, 80 (1984) 629.

583g Catalytic Deoxygenation of Stearic and Oleic Acids Over Pd/C

Jeremy G. Immer[1], M. Jason Kelly[1], Benjamin D. Estabrook[2], and H. Henry Lamb[3]. (1) Chemical and Biomolecular Engineering, North Carolina State University, 911 Partners Way, Raleigh, NC 27695, (2) Middlebury University, 2885 Middlebury College, Middlebury, VT 05753, (3) NC State University, Chemical Engineering, Raleigh, NC 27695-7905

Catalytic deoxygenation of C18 free fatty acids (FFAs) can be used to produce C17 hydrocarbons as an alternative to conventional fatty acid methyl ester (FAME) biodiesel [1]. FFA deoxygenation occurs via two pathways: decarbonylation and decarboxylation. In this work, deoxygenation of oleic and stearic acids over a 5 wt.% Pd/C catalyst was investigated to determine the effects of unsaturation of the alkyl chain and the feasibility of using the hydrocarbon reaction product as solvent. FFA deoxygenation was conducted in a semi-batch autoclave at 300ºC and 15 atm using either n-dodecane or n-heptadecane as solvent with a constant flow of 60 mL/min of He or 10% H2 in He. Deoxygenation of stearic acid occurs primarily via a decarboxylation pathway in either He or 10% H2 flow. The initial decarboxylation rate under He is much higher than under 10% H2, but the reaction reaches completion more quickly in the latter case because H2 inhibits catalyst deactivation. Deoxygenation of oleic acid is strongly inhibited under He, achieving a conversion of only 12% after 3 h with less than 10% selectivity to heptadecane. In contrast, the oleic acid deoxygenation goes to completion in <1 h in the presence of H2 yielding n-heptadecane in essentially 100% yield. Additional experiments established that the overall reaction for oleic acid is a stepwise process: initial hydrogenation of the double bond to form stearic acid and subsequent decarboxylation to n-heptadecane. Using n-heptadecane as solvent, the initial rate of stearic acid decarbonylation under 10% H2 was equivalent to that observed in n dodecane. The initial rate of decarboxylation, however, was substantially lower due to reversibility making decarbonylation a more significant pathway in the overall reaction. Stearic acid deoxygenation nears completion in n-heptadecane in ~2 h aided by removal of the CO2 reaction product.

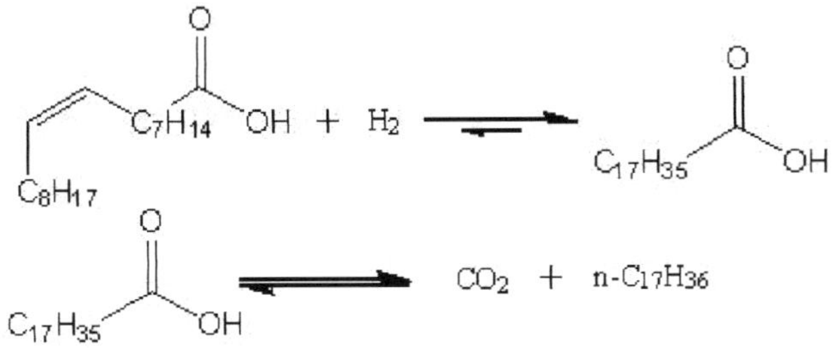

Catalytic Deoxygenation of Liquid Biomass for Hydrocarbon Fuels

Howard Mayfield, Ted Marolla, and Bob Nichols, AFRL, Tyndall Air Force
Mike Mashburn, Physics Department, Florida State University
Kyung C Kwon, Chemical Engineering Department, Tuskegee University,Tuskegee, AL 36088, Tel. 334-727-8976/724-4528, kwonk@tuskegee.edu

Abstract

Neat methyl laurate was chosen as a model compound of liquid biomass for this research. Removal of oxygen from liquid biomass is a key step in the elevation of their energy intensities. Canola oil and its biodiesel, spent waste kitchen oil and its biodiesel, NSWC-02 biodiesel (6/13/08), Energy BLD 9738 (1/8/08), and Wallace biodiesel (7/14/2008), in addition to methyl laurate were deoxygenated with hydrogen and a catalyst.

Oxygen from methyl laurate and liquid biomass is removed at 250 – 1220 psig initial cold hydrogen pressure and 300 – 400°C, using a 316 stainless steel batch reactor. Removal of oxygen from liquid methyl laurate and other biomass is evaluated with a GC/MS analyzer for liquid reaction products, while consumption of hydrogen and conversion of liquid biomass to gaseous reaction products are analyzed with another GC. The range of reaction duration of liquid biomass in the batch reactor is 30 – 60 minutes under the deoxygenation conditions. Consumption of hydrogen and conversion of liquid biomass into gaseous products are evaluated with analyzing gaseous reaction products and making a mass balance on hydrogen.

The objective of this research is to develop an efficient method of removing oxygen from liquid biomass to produce liquid petroleum-comparative hydrocarbons with the following specific aims:

1. Formulate catalysts for the removal of oxygen from liquid biomass.
2. Test capabilities of formulated catalysts for the deoxygenation and isomerization of liquid biomass.
3. Identify reaction conditions suitable for the conversion of liquid biomass into liquid hydrocarbon fuels
4. Determine reaction mechanisms on the removal of oxygen compounds from liquid biomass.
5. Develop an economic method of regenerating catalysts deactivated with carbonaceous deposits and sulfur loss.

Introduction

Liquid biomass such as methyl laurate, methyl stearate, canola oil and others are chosen to study deoxygenation and isomerization of liquid biomass. Removal of oxygen from liquid biomass for renewable energies and isomerization of liquid biomass are needed to increase its energy intensity, stability, and decrease its viscosity. Liquid biomass will be deoxygenated with hydrogen and active catalytic metal candidates such as Ni, Mo, Pd, Co, and Zn. Liquid biomass will be isomerized with hydrogen and Al_2O_3 or SiO_2. These metals will be impregnated into a γ-alumina support.

Conversion of liquid biomass to hydrocarbon fuels is the primary step in refining liquid biomass produced from vegetable oils, animal fats, biodiesels, and waste cooking oils. Removal of oxygen from biofuels is needed to enhance fuel enthalpies, decrease viscosity of biomass, and stabilize biofuels. Current method of deoxygenating liquid biomass with pellet catalysts is equipment intensive and expensive, requiring a packed beds reactor, a fluidized beds reactor, and a trickle bed reactor.

Many researchers are developing the Catalytic Conversion Process (CCP) of liquid biomass for hydrocarbons, using catalyst pellet particles and hydrogen in a packed-bed reactor, a fluidized bed reactor, and a trickle bed reactor to deoxygenate and isomerize liquid biomass. However, significant pressure drops and non-uniform flow of liquid biomass and gaseous hydrogen were experienced over a packed-bed reactor. Tuskegee University undertakes a feasibility study on replacing catalysts in fine pellet form with monolithic form for the CCP process.

Hydrogen and liquid biomass are fed downward into a vertical monolithic catalyst reactor at 250 to 450^oC in the CCP process for the conversion of liquid biomass into liquid hydrocarbons to enhance heating values of biofuels, stabilize biofuels, and decrease viscosity of biofuels. In the CCP process, liquid biomass and hydrogen can be mixed and converted uniformly with honeycomb monolithic catalysts in comparison with a pellet-packed reactor. Regeneration of monolithic catalyst is simple in comparison with pellet-type catalysts

Various catalysts in addition to NiMo/γ-Al_2O_3 and CoMo/γ-Al_2O_3 are well known for deoxygenating liquid biomass; however, these catalysts were formulated and tested mostly for refining and hydro-cracking petroleum products. Research is needed in the development of dual function catalysts suitable for both deoxygenizing and isomerizing functions to advance the CCP process, in which liquid hydrocarbons are formed. In this paper, preliminary results of experiments on removal of oxygen from liquid biomass, using a batch reactor, a GC/MS, and a GC, are described.

Objectives

The objective of this research is to develop an efficient removal method of oxygen from biofuels to minimize oxygen contents in biofuels with suitable catalysts with the following specific aims:

1. Formulate catalysts for the removal of oxygen from liquid biomass.
2. Test capabilities of formulated catalysts for the deoxygenation and isomerization of liquid biomass.
3. Identify reaction conditions suitable for the conversion of liquid biomass into liquid hydrocarbon fuels
4. Determine reaction mechanisms on the removal of oxygen compounds from liquid biomass.
5. Develop an economic method of regenerating catalysts deactivated with carbonous deposits and sulfur loss.

Biomass such as methyl laurate, methyl stearate, canola oil, and others are chosen as a liquid biomass for this preliminary research. Removal of oxygen from liquid biomass and isomerization of liquid biomass are needed to increase its energy intensity, stabilize it, and decrease its viscosity. Oxygen from liquid biomass is removed and liquid biomass is isomerized at 100 – 1250 psig initial cold hydrogen pressure and 250 – 450°C, using a batch reactor. Removal of oxygen from liquid biomass and consumption of hydrogen is analyzed with a GC/MS for liquid reaction products. Gaseous reaction products were analyzed with a GC. The reaction duration range of liquid biomass in a monolithic catalyst flow reactor is 10 – 60 minutes under the deoxygenating conditions. Consumption of hydrogen and conversion of liquid biomass into gaseous products are evaluated by analyzing gaseous reaction products with the GC and making a mass balance on hydrogen.

Experimental Setup

Oxygen from methyl laurate and biomass is removed at 250 – 1220 psig and 300 – 400°C, using a 35-cm^3 batch reactor made of 316 stainless steel. Removal of oxygen from liquid methyl laurate and other biomass is evaluated with a GC/MS analyzer for liquid reaction products and gaseous reaction products are analyzed with another GC. The range of reaction duration of liquid biomass in the batch reactor is 30 – 60 minutes under the deoxygenating conditions.

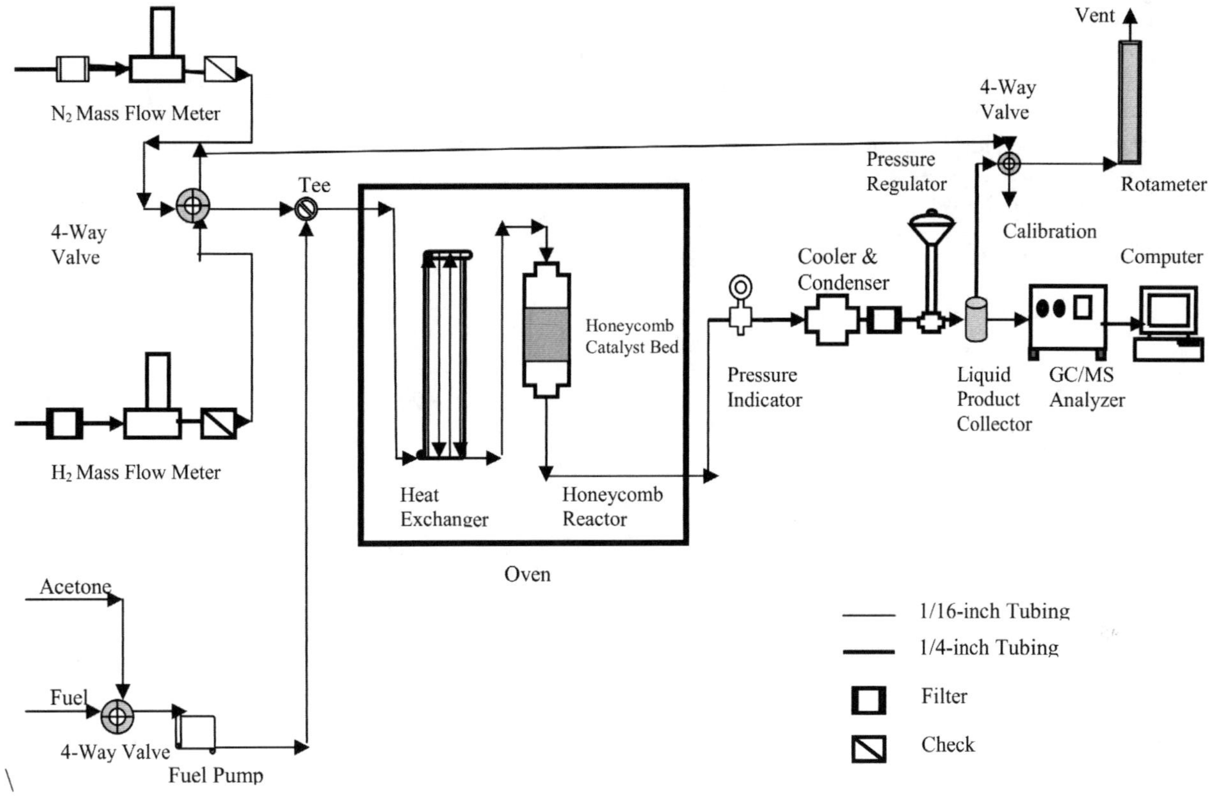

Figure 1. Schematic Diagram on a deoxygenation reactor assembly

Typical Reaction Conditions

- Batch Reactor Volume, cm3: 35
- Temperature, oC 300 – 400
- Initial Cold Hydrogen Pressure, psig: 250 – 1220
- Amount of Biomass, g: 2
- Amount of Catalyst, g: 0 – 0.2
- Biomass:

Methyl Laurate
Canola Oil and its Biodiesel
Spent Waste Kitchen Oil and its Biodiesel
NSWC-02 Biodiesel (6/13/08)
Energy BLD 9738 Biodiesel (1/8/08)
Wallace Biodiesel (7/14/08)

Equipment

- GC/MS for Liquid Reaction Products:
- GC for Gas Products:
- Batch Reactor
- Heater and its Temperature Controller
- Wrist-Arm Shaker and Horizontally-Moving Orbital Shaker

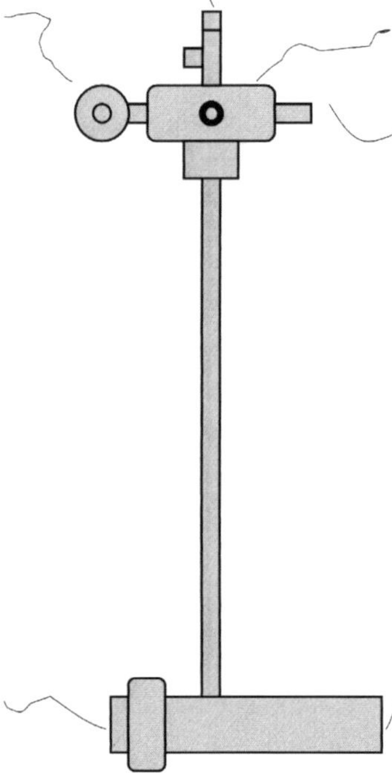

Figure 2. High-pressure and high-temperature batch reactor

Reaction Pathways

Biomass → Biodiesel

Biodiesel → $H_2O + CO_2 + CH_4$ + Hydrocarbons

Biomass → $H_2O + CO_2 + CH_4$ + Hydrocarbons

Conversion of Methyl Laurate into Dodecane

$$CH_3(CH_2)_{10}COOCH_3 + 3H_2 \rightarrow CH_3(CH_2)_{10}CH_3 + H_2O + CH_4$$

Conversion of Methyl Laurate into Undecane

$$CH_3(CH_2)_{10}COOCH_3 + H_2 \rightarrow CH_3(CH_2)_9CH_3 + CO_2 + CH_4$$

Results

GC/MS Analysis: Identification/Quantitation Standard

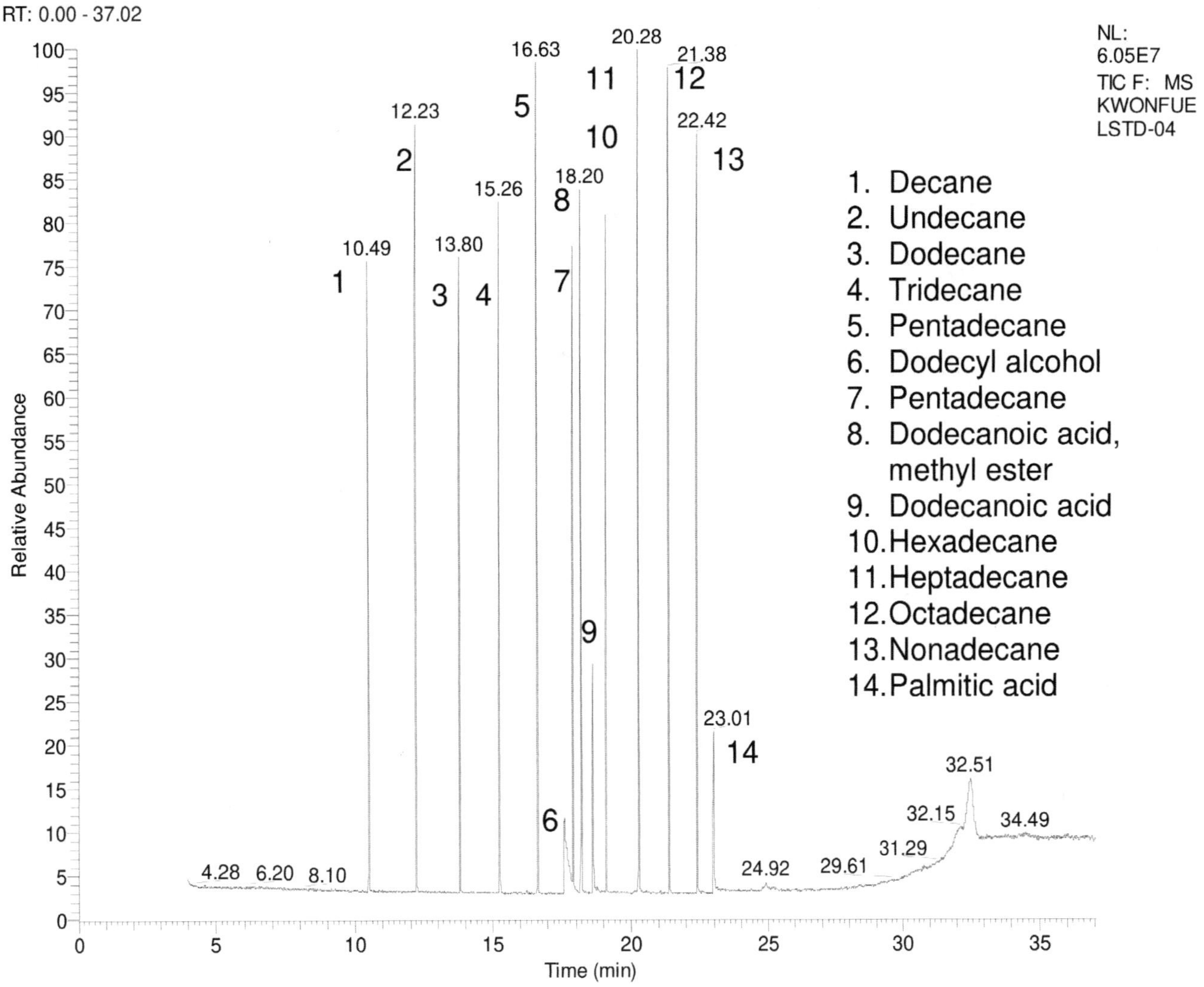

Catalytic Hydrogenation of Canola Oil

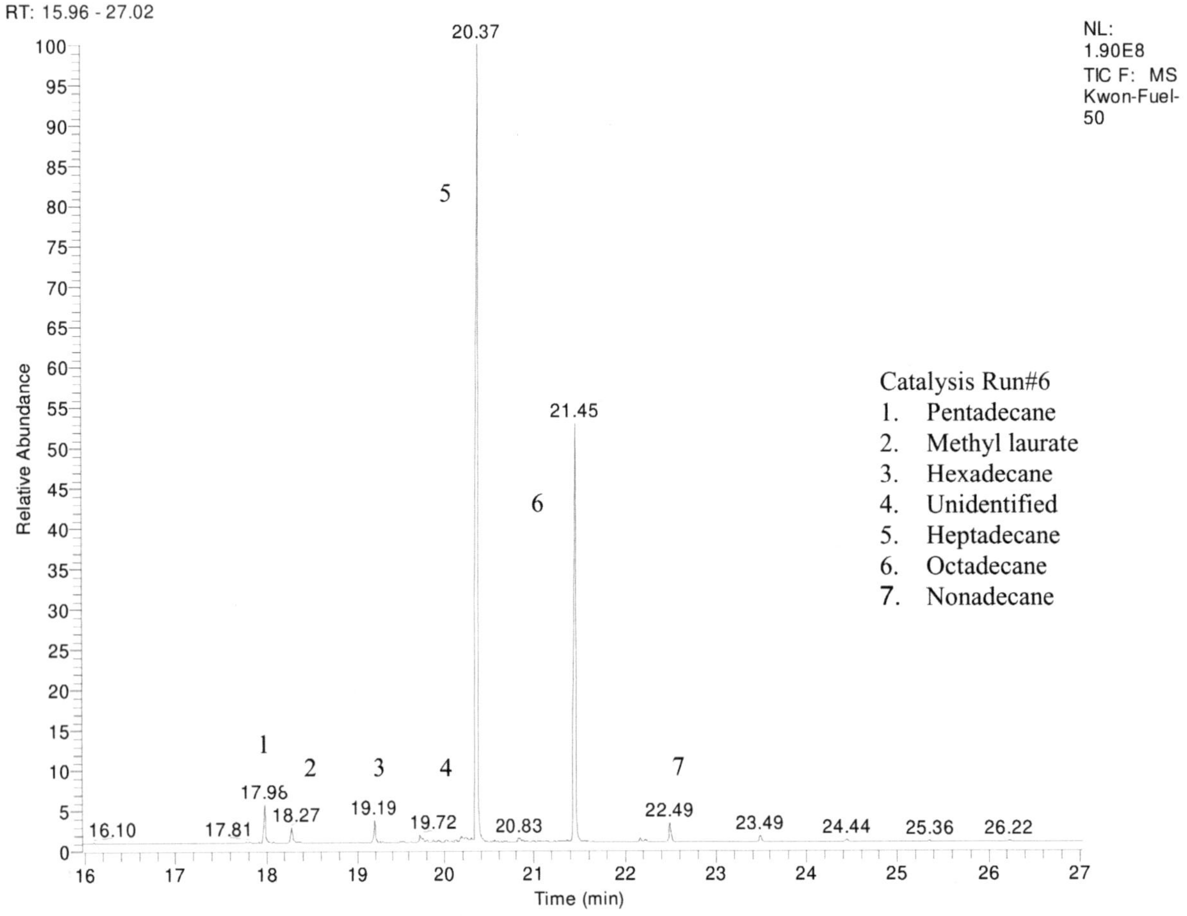

Catalytic Hydrogenation of Methyl Laurate

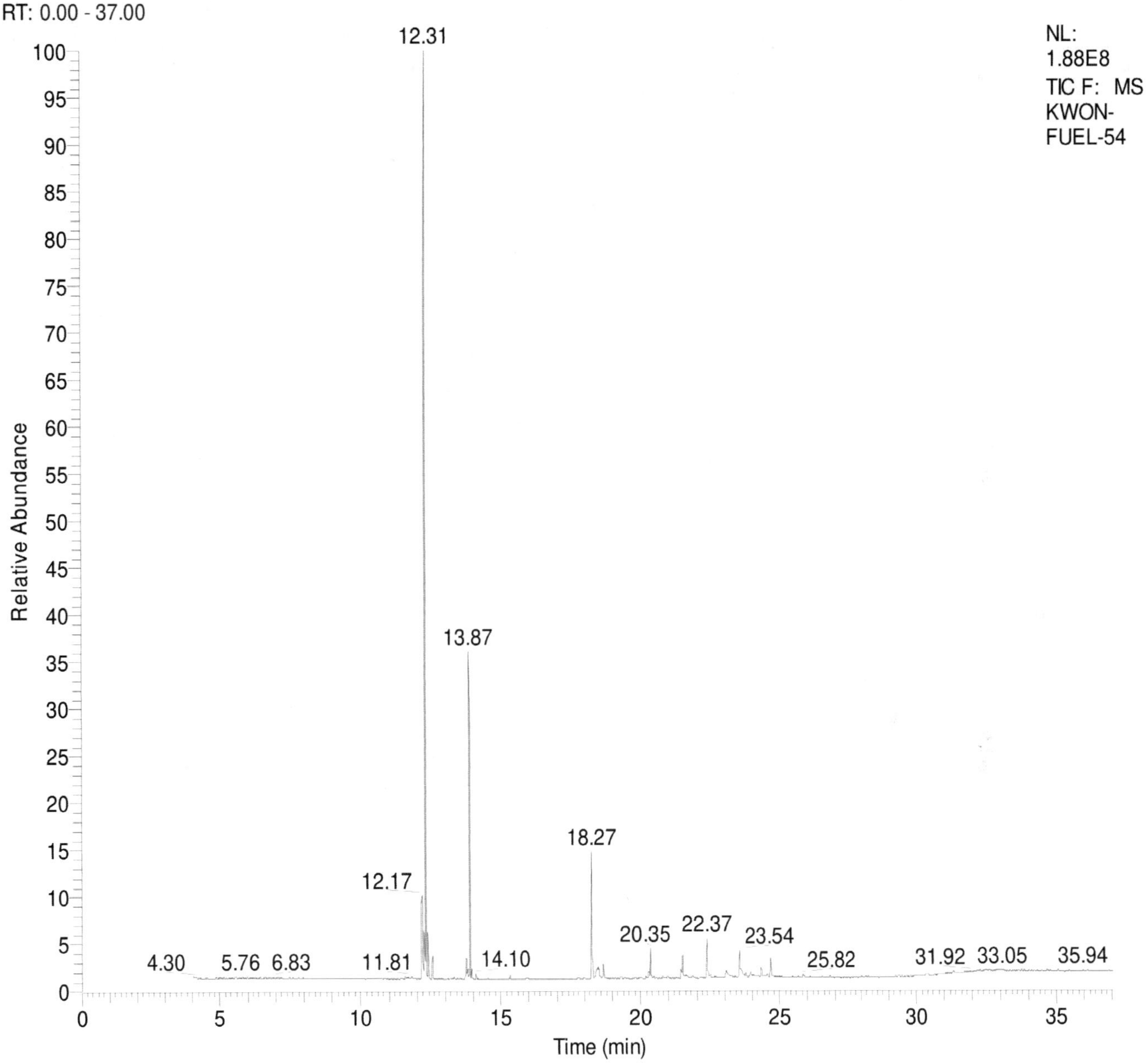

Table 1. Results on catalytic deoxidizing biomass at 300 – 400 °C and 250 - 1220 psig cold initial hydrogen pressure

	Run #	1	2	3	4	5	6	7	8	9	10	11	12	13
	Date	7/9/08	7/9/08	7/10/08	7/11/08	7/11/08	7/11/08	7/11/08	7/14/08	7/14/08	7/15/08	7/15/08	7/16/08	7/16/08
	Temp, oC	350	350	350	350	350	350	350	350	350	350	350	350	350
	Reaction Duration, min	30	30	30	30	30	30	30	30	30	30	30	30	30
Reactor Charge, g	NiMo catalyst	0.101	0.1038	0.1035	0.1010	0	0.1032	0.1003	0.1022	0.0512	0.0258	0.104	0.1259	0.1026
	Methyl Laurate	3.0200	1.8618	2.0011	1.9605	2.0128		1.9997	1.9325	2.0126	1.9935	1.971	1.9895	
	Canola Oil						2.0157							
	Spent Waste Kitchen Oil													1.9992
	Decane	6.863	3.0004											
H_2 Press, psig	t=0 min, T=25°C	na	na	680	330	670	650	520	670	670	660	660	680	670
	t=30 min, T=25°C	na	na	610	340	670	470	480	670	600	640	600	580	590
	t=30 min, T=T°C	na	na	1190	720	1100	840	930	1080	1130	1120	1120	1130	970
Liquid Products, Weight %														
	decane	73.07	53.80											
	undecane	3.22	2.29	39.03	33.38	1.81		40.17	0.91	41.58	11.20	55.17	60.65	8.10
	dodecane	1.54	1.66	28.84	10.49	0.51		18.20	0.21	24.06	4.15	31.73	34.67	
	dodecanol	0.00	11.99	8.37										
	methyl_laurate	20.89	24.28	9.08	8.16	95.22	1.58	6.27	75.44	12.48	25.64	6.45	0.79	5.40
	lauric_acid	1.69	2.34	1.32	2.67	2.35		2.60	21.44	4.49	6.81			
	hexadecane						1.82							
	heptadecane						57.23	1.13						44.57
	octadecane				3.62		34.50	1.64						20.26
	nonadecane				5.24		1.47	2.20		1.04				
	heptadecanoic_acid							3.53						
	unaccountable	-0.41	3.59	13.16	35.84	0	3.41	23.78	1.02	15.49	50.82	6.43	3.49	21.67
Gaseous Products	H_2 V % in gas product	na	na	na	48.95	98.28	43.59	49.41	96.46	78.82	90.4	79.3	77.88	95.65
	CH_4 V % in gas product	na	na	na	0	0	0	28.74	3.29	18.2	10.2	20.6	16.06	0
	H_2 Initial Amount , g	na	na	0.1363	0.0676	0.1343	0.1304	0.1049	0.1343	0.1343	0.1324	0.1324	0.1363	0.1343
	H_2 Consumption, g	na	na	na	0.0336	0.0023	0.0890	0.0569	0.0048	0.0393	0.0163	0.0367	0.0454	0.0209
	H_2 Consumption, w %, basis on initial H_2	na	na	na	49.63	1.72	68.214	54.286	3.54	29.24	12.28	27.75	33.3306	15.5257
	H_2 Consumption, %, basis on initial biomass	na	na	na	1.7120	0.1148	4.4133	2.8479	0.2461	1.9516	0.8154	1.8637	2.2834	1.0432
	CH_4 Production, g	na	na	na	0.0000	0.0000	0.0000	0.2219	0.0352	0.1746	0.1042	0.1977	0.1491	0.0000
	CH_4 Production, w %, basis on initial biomass	na	na	na	0.0000	0.0000	0.0000	11.099	1.8196	8.6772	5.2291	10.0032	7.4938	0.0000

Table 2. Results on catalytic deoxidizing biomass at 300 – 400 °C and 250 - 1220 psig cold initial hydrogen pressure

	Run #	14	15	16	17	18	19	20	21	22	23	24	25
	Date	7/21/08	7/21/08	7/21/08	7/21/08	7/22/08	7/22/08	7/22/08	7/23/08	7/23/08	7/23/08	7/23/08	7/23/08
	Temp, oC	350	350	300	325	375	375	400	350	350	350	350	350
	Reaction Duration, min	30	30	30	30	30	30	30	30	30	30	30	30
Reactor Charge, g	NiMo catalyst	0.102	0.0267	0.1053	0.1022	0.1008	0.1006	0.1008	0.1005	0.1006	0.2002	0.026	0.0505
	Methyl Laurate	1.9445	1.9986	2.0836	2.0003	1.9861			0.0000	0.0000	0	0	0
	Canola Oil						2.0303	1.9974	0.5012	1.006		2.044	2.0183
	Spent Waste Kitchen Oil								1.4912	1.036	1.9978	0	0
	Decane												
H_2 Press, psig	t=0 min, T=25°C	360	670	670	660	660	660	660	660	670	670	670	670
	t=30 min, T=25°C	370	670	590	530	610	570	520	570	590	470	410	525
	t=30 min, T=T°C	590	1110	940	930	1190	980	1020	980	1015	820	700	900
Liquid Products, Weight %		Run 14	Run 15	Run 16	Run 17	Run 18	Run 19	Run 20	Run 21	Run 22	Run 23	Run 24	Run 25
	decane												
	undecane	5.45	5.20	1.49	3.54	42.18							
	dodecane	1.99	2.07	1.44	2.33	17.19							
	dodecanol												
	methyl_laurate	51.67	68.45	82.78	67.93	7.92	5.65	8.83	3.80	3.01	3.50	1.01	1.46
	lauric_acid	24.87	14.63	12.05	19.47	4.74							
	hexadecane						1.61	2.33	1.39	1.28	1.42		
	heptadecane						27.78	33.19	20.30	19.37	14.82	8.80	14.20
	octadecane						10.63	11.78	10.21	11.30	8.13	2.39	4.61
	nonadecane												
	heptadecanoic_acid						9.69	6.39	8.10	10.43	16.16	8.29	6.93
	unaccountable	15.50	9.13	1.98	6.46	27.70	40.01	35.23	55.42	53.86	55.33	78.98	71.88
Gaseous Products	H_2 V % in gas product	89	98.84	99.86	96.6	86	96.67	92.83	98.3	99.5	95	98.61	96.34
	CH_4 V % in gas product	11.3	4.18	3.21	5.9	12.34	0	0	0	0	0	0	0
	H_2 Initial Amount , g	0.0735	0.1343	0.1343	0.1324	0.1324	0.1324	0.1324	0.1324	0.1343	0.1343	0.1343	0.1343
	H_2 Consumption, g	0.0063	0.0016	0.0159	0.0291	0.0270	0.0215	0.0350	0.0196	0.0163	0.0440	0.0522	0.0323
	H_2 Consumption, w %, basis on initial H_2	8.625	1.16	11.808	22.013	20.373	16.225	26.432	14.812	12.13	32.749	38.84	24.0621
	H_2 Consumption, %, basis on initial biomass	0.3261	0.0780	0.7613	1.4567	1.3579	1.0579	1.7517	0.9841	0.7978	2.2021	2.5525	1.6015
	CH_4 Production, g	0.0679	0.0447	0.0303	0.0502	0.1203	0.0000	0.0000	0.0000	0.0000	0.0000	0.0000	0.0000
	CH_4 Production, w %, basis on initial biomass	3.4898	2.2354	1.4542	2.5079	6.0588	0.0000	0.000	0.0000	0.0000	0.0000	0.0000	0.0000
Gas Product mass, g	Total	0.404	0.3883	0.3442	0.2657	0.4913	0.4692	0.419	0.3785	0.344	0.3687	0.245	0.34234
	H_2	0.067	0.1328	0.1185	0.1032	0.1054	0.1109	0.0974	0.1128	0.118	0.0903	0.082	0.10201
	Others	0.337	0.2555	0.2257	0.1625	0.3859	0.3583	0.3216	0.2657	0.226	0.2784	0.163	0.24033
Gasified Biomass, w %, basis on initial biomass		17.34	12.78	10.83	8.12	19.43	17.65	16.10	13.34	11.07	13.93	7.96	11.91

Table 3. Results on catalytic deoxidizing biomass at 300 – 400 °C and 250 - 1220 psig cold initial hydrogen pressure

	Run #	26	27	28	29	30	31	32	33	34	35	36	37
	Date	7/24/08	7/24/08	7/24/08	7/24/08	7/24/08	7/25/08	7/25/08	7/29/08	7/29/08	7/29/08	7/29/08	7/29/08
	Temp, oC	350	350	350	350	350	350	350	350	350	350	350	350
	Reaction Duration, min	30	30	60	30	60	30	30	30	30	30	30	
Reactor Charge, g	NiMo catalyst	0.076	0.1016	0.1	0.1002	0.1015	0.1004	0.0999	0.1007	0.1017	0.1002	0.102	0.1002
	Methyl Laurate	0	0	0	0	0	1.9797	0					
	Canola Oil	2.042	1.9954	2.0042	2.0005	2.0379	0	2.0145					
	Spent Waste Kitchen Oil												
	canola biodiesel								2.0448				
	spent Oil biodiesel									2.006			
	NSWC-02 (6/13/08)										2.0468		
	Energy BLD 9738 (1/8/08)											2.008	
	wallace biodiesel 7/13/08												2.035
	Decane												
H_2 Press, psig	t=0 min, T=25°C	670	670	660	250	1220	660	670	660	660	660	650	660
	t=30 min, T=25°C	620	490	185	220	500	610	580	585	580	650	590	610
	t=30 min, T=T°C	1020	870	370	350	940	1120	1020	1020	980	1060	1000	1040
	decane												
	undecane						14.41				1.60		
	dodecane						12.32				1.49		
	dodecanol											3.19	1.38
	pentadecane				1.45	4.76		1.70	1.25	1.66	4.52		
	methyl_laurate	1.85	2.10	1.82			57.58	3.72			2.59		
	lauric_acid						8.51						
	hexadecane					1.75					2.20		
	heptadecane	19.64	20.58	16.18	17.07	46.97		17.15	12.98	8.08	21.66	6.40	4.77
	octadecane	6.22	6.54	3.78	4.85	23.35		10.20	4.64	2.21	14.16	1.88	1.33
	nonadecane					1.70			2.40		3.47	6.64	7.56
	heptadecanoic_acid	4.86	5.31	5.18	2.77			3.63		2.81		3.62	5.33
	unaccountable	66.24	64.03	71.80	72.73	21.20	5.46	60.76	77.52	84.72	46.43	77.56	78.99
Gaseous Products	H_2 V % in gas product	99.67	97.87	81.12	91.37	90.42	91.31	95.37	92.32	94.52	96.14	95.11	97
	CH_4 V % in gas product	0.00	0.00	0.00	0.00	0.00	7.86	0.00	7.39	6.09	2.93	4.99	3.93
	H_2 Initial Amount , g	0.1343	0.1343	0.1324	0.0519	0.2422	0.1324	0.1343	0.1324	0.1324	0.1324	0.1304	0.1324
	H_2 Consumption, g	0.0102	0.0374	0.1006	0.0099	0.1509	0.0205	0.0231	0.0238	0.0221	0.0070	0.0176	0.0135
	H_2 Consumption, w %, basis on initial H_2	7.608	27.856	75.989	18.982	62.306	15.456	17.167	17.942	16.69	5.2849	13.48	10.1884
	H_2 Consumption, %, basis on initial biomass	0.5005	1.8753	5.0189	0.4928	7.4062	1.0335	1.1447	1.1615	1.1011	0.3418	0.8754	0.6627
	CH_4 Production, g	0.0000	0.0000	0.0000	0.0000	0.0000	0.0767	0.0000	0.0692	0.0565	0.0304	0.0471	0.0383
	CH_4 Production, w %, basis on initial biomass	0.000	0.000	0.000	0.000	0.000	3.874	0.000	3.383	2.818	1.485	2.346	1.883
Gas Product mass, g	Total	0.338	0.334	0.291	0.2989	0.4533	0.4006	0.4067	0.330	0.407	0.423	0.435	0.401
	H_2	0.124	0.097	0.032	0.0421	0.0913	0.1119	0.1113	0.109	0.110	0.125	0.113	0.119
	Others	0.214	0.237	0.259	0.2568	0.362	0.2886	0.2954	0.221	0.296	0.297	0.322	0.282
Gasified Biomass, w %, basis on initial biomass		10.48	11.90	12.93	12.84	17.76	14.58	14.66	10.83	14.77	14.52	16.04	13.84
Remarks													

634a Gas Phase Processing of Sorbitol

Galen J. Suppes and Wei Yan. University of Missouri-Columbia, Department of Chemical Engineering, W2028 Engineering Bldg. East, Columbia, MO 65211

During the conversion of glycerol to propylene glycol, highly selective conversion is necessary for commercial viability. The greatest strides in achieving high selectivity are attained with catalyst and temperature. For the conversion of glycerol to propylene glycol, these parameters can be optimized to achieve selectivities of greater than 80%.

This paper is on the optimization of more-subtle parameters such as concentration, water content, pressure, isothermal operation, and residence time to achieve selectivities in excess of 90%. Data reveal that low concentrations are important to reduce by-products whose formation relies on second-order reaction mechanisms. Water is important to reduce dehydration reactions and indirectly helps to maintain more-isothermal operation. An optimal hydrogen partial pressure between 5 and 15 bars minimizes the cumulative amount of by-products that result from hydrocracking versus dehydration side reactions.

634b Conversion of Biomass-Derived Polyols Over Carbon Supported Pt-Re Catalysts

Edward L. Kunkes, Dante A. Simonetti, Ryan M. West, and James A. Dumesic. Department of Chemical and Biological Engineering, University of Wisconsin - Madison, 1415 Engineering Drive, Madison, WI 53706

We have studied the conversion of concentrated polyol solutions (80 wt% glycerol and 60 wt % sorbitol) into gas mixtures containing CO/CO2/H2 and light alkanes as well as oxygenated hydrocarbons (e.g., alcohols and ketones) in the liquid phase over a Pt-Re/C catalyst. The conversion of polyols over Pt-Re/C proceeds via reaction pathways that involve C-C and C-O bond scission. High rates of C-C scission lead to the formation of CO and CO2, while high rates of C-O scission result in alkanes, alcohols, diols, and ketones. We show that the relative rates of C-C versus C-O scission can be manipulated by changing process conditions (temperature and pressure). At 503 K and 18 bar, the conversion of 80 wt % glycerol yields COx (40% carbon conversion), C1-C3 alkanes (20%), acetone (5%), and alcohols and diols such as ethanol, ethylene glycol, and acetol (35%). Increasing the temperature to 523 K at constant pressure causes the selectivity to COx to increase to 50% and alkane selectivity to increase to 27%, while the selectivity to alcohols decreases to 18%. Increasing the pressure to 27 bar at constant temperature causes a significant increase in the production of alcohols (50% selectivity) at the expense of alkanes (13%) and COx (30%). We also show that under similar reaction conditions, the conversion of glycerol produces more COx and light alcohol species, whereas sorbitol yields more alkanes and higher molecular weight oxygenates, and that both polyols follow similar pressure and temperature trends with respect to relative rates of C-C to C-O bond cleavage. The results from this study provide a fundamental basis for using Pt-Re catalysts to convert polyols to fuels and valuable chemicals.

CATALYTIC DEOXYGENATION OF BENZALDEHYDE OVER SUPPORTED GALLIUM CATALYSTS

Artit Ausavasukhi[a], Daniel E. Resasco[b], and Tawan Sooknoi[a]
[a] *King Mongkut's Institute of Technology Ladkrabang, BKK, Thailand.*
[b] *University of Oklahoma, Norman, OK, USA.*

Abstract

The deoxygenation of benzaldehyde was investigated over supported Ga catalysts as part of a program leading to the study of model reactions involving aldehydes, acids, furfurals, and oxygenated aromatics that may be present in pyrolysis oil (bio-oil). Ga/SiO_2 catalysts were prepared by impregnating precipitated SiO_2 supports with $Ga(NO_3)_3$ solution. The reaction conditions, H_2 concentration, Ga content, and Ga dispersion in the catalysts were varied to investigate the effect of these parameters on activity, selectivity, and stability. TPR gave evidence for a partial reduction of Ga/SiO_2 in the range 500-750 °C. Increasing H_2 concentration (3-100%v/v) was found to enhance activity (~50-90%) and toluene selectivity (40-75%). Reaction at low temperature (< 500 °C) leads to the formation of high molecular weight oxygenates and consequently rapid catalyst deactivation (< 2 h). In contrast, no benzaldehyde conversion can be observed over a physical mixture of Ga_2O_3 and SiO_2.

Introduction

Bio-oils derived from biomass have been increasingly attractive as alternative sources of fuels and chemicals. This is due to an increase in petroleum prices, government regulations on alternative energy, and commitments to greenhouse gas reduction. However, the bio-oils contains various types of oxygenates that can cause problems when used as fuels. Among these, aldehyde and carboxylic are the most unwanted components. Deoxygenation of bio-oils does not only improve it fuel properties, but also provide alternative sources of petrochemicals.

This research deals with the conversion of benzaldehyde, a model compounds representing aromatic aldehydes in bio-oil, to oxygen-free compounds. It was found that nickel [1], platinum [2], palladium [3], and copper [4] based catalysts are readily active for hydrogenation of the carbonyl compound. In addition to noble metal, Ga based catalyst is proposed for the deoxygenation of benzaldehyde to produce oxygen-free aromatics. The study on catalytic performance of Ga support SiO_2 involves effect of reaction conditions, H_2 concentration, Ga content, and Ga dispersion in the catalysts.

Experimental Details

Catalyst preparation

Spherical porous SiO_2 support was prepared by precipitation of sodium silicate solution with sodium hydroxide [5-6]. Ga supported SiO_2 (1-6 wt%) was prepared by impregnation of the SiO_2 support with $Ga(NO_3)_3$ (%loading-Ga/SiO_2), and physical mixed of the support with bulk Ga_2O_3 (*b-*

Ga/SiO_2). All the samples were then activated in air at 550 °C for 4 hours. The specific surface area measurements were performed on an Autosorb-1 using N_2 as a probe molecule using BET method

Temperature-programmed reduction

Temperature-programmed reduction (TPR) experiments were carried out using a TCD detector. The sample was treated at 550 °C for 1 hour in air and then cooled to room temperature under He flow. The flow was then switched to 10% H_2/Ar (30 ml/min) and the sample was heated up to 900 °C at a rate of 10 °C/min.

Catalytic testing

Catalytic testing of benzaldehyde conversion was performed with a continuous fixed bed reactor. Saturated vapor of benzaldehyde at 20 °C was carried by H_2 (or H_2/He mixture) at 30 mL/minute regulated by a mass flow controller. The catalyst bed was set in a quartz tube reactor (ID 6 mm) that was located inside a temperature-regulated furnace. The products were periodically collected and analyzed by online gas chromatography (GC). The preferred reaction conditions used in the experiments were as follows: temperature, 450-550 °C; total pressure, 1 atm; carrier gas, He/H_2; W/F, 100-375 g·h/mol.

Results and discussion

Catalyst characterization

The specific surface area of supported Ga catalysts both 3Ga/SiO_2 (325 m^2/g) and *b*-3Ga/SiO_2 (355 m^2/g) are slightly lower than that of the SiO_2 support (370 m^2/g). This may be derived from a surface reaction of the loaded Ga with the support. However, all samples still possess high surface area.

Catalytic activity

Benzaldehyde conversion

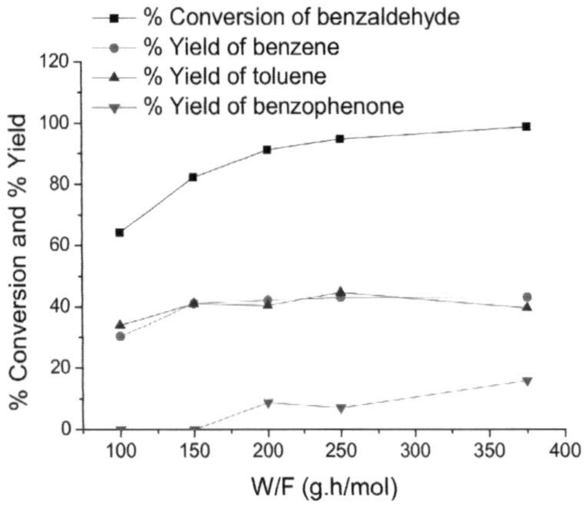

Fig. 1 Effect of W/F of benzaldehyde conversion over 3Ga/SiO_2.

(W/F = 100-375 g.h/mol, reaction temperature 550 °C, 100 % H_2 concentration, total flow 30 mL/min, the data was shown at initial time on stream)

From Fig.1, it is found that benzaldehyde can be converted to benzene, toluene and benzophenone. Moreover, such products increase with W/F. It is expected that the formation of benzene and toluene are taken place from the parallel reaction, possibly derived from the coupled precursor, namely benzophenone.

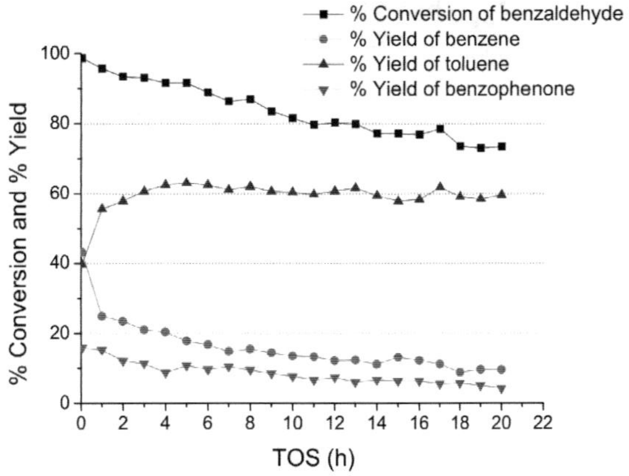

Fig. 2 Deactivation profile of benzaldehyde conversion over 3Ga/SiO_2 catalyst.

(W/F = 375 g.h/mol, reaction temperature 550 °C, 100 % H_2 concentration, total flow 30 mL/min)

From Fig. 2, change in product yield with time on stream can be observed along with the continual deactivation of the catalyst. This is the case for all W/F tests. Benzene and toluene are equally produced at the beginning of the reaction. Interestingly, the toluene product is initially increased with time on stream, while benzene and benzophenone are gradually reduced.

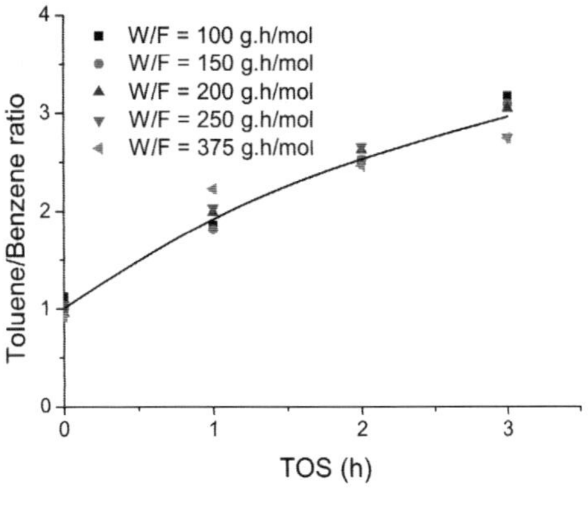

Fig. 3 Toluene/benzene ratio at different time on stream over 3Ga/SiO_2.

(W/F = 100-375 g.h/mol, reaction temperature 550 °C, 100 % H_2 concentration, total flow 30 mL/min)

For all W/F tests, the toluene/benzene ratio is initially unity and gradually increased with time on stream (Fig. 3). This is presumably explained that benzaldehyde can be readily coupled to form aromatic diketone [7]. This intermediate consecutively undergoes ring contraction liberating CO (from a carbonyl group) to form benzophenone. This benzophenone can be hydrogenolysed to form benzene and toluene as observed. At initial conversion, the decarbonylation activity of the diketone was active to generate benzophenone and readily cleave to form equal molar of benzene and toluene. However, as the catalyst is deactivated, aromatic diketone is accumulated on the surface. This diketone can be directly hydrogenolysed to toluene. This leads to the observed increase in toluene/benzene ratio with time on stream. The schematic diagram shows the formation of product proposed. (Scheme I)

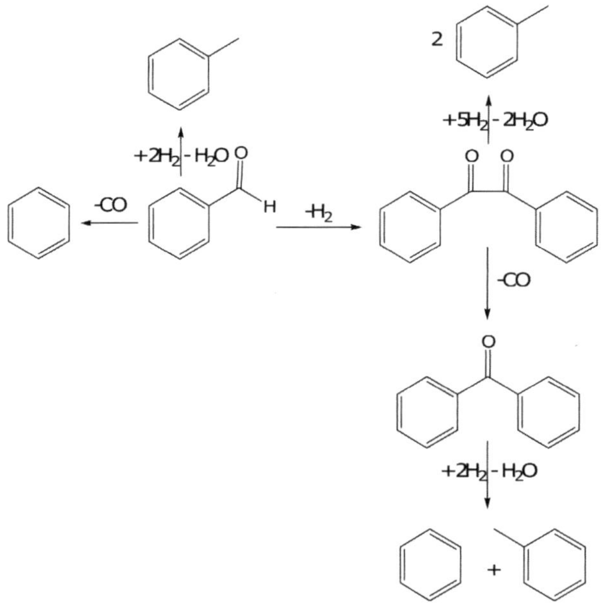

Scheme I

Effect of Temperature

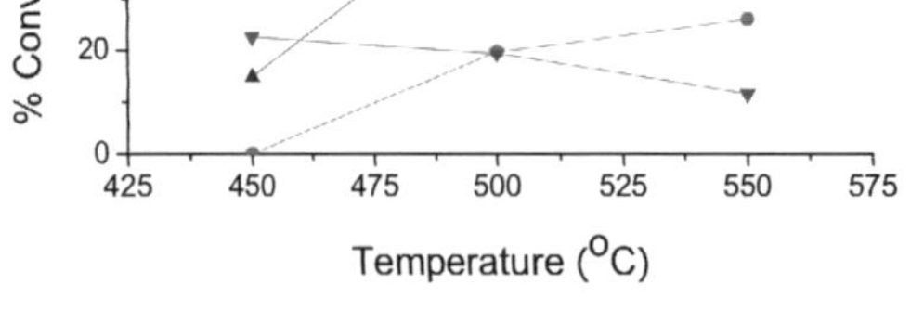

Fig. 4 The effect of reaction temperature on benzaldehyde conversion over $3Ga/SiO_2$

(W/F = 250 g.h/mol, Reaction temperature 450-550 °C, 100 % H_2 concentration, Total flow 30 mL/min, the data was shown at 1 hour on stream)

From Fig.4, it is found that reaction at low temperature (450 °C) gives a low benzaldehyde conversion. A higher benzaldehyde conversion was rapidly increased, when the temperature was risen.

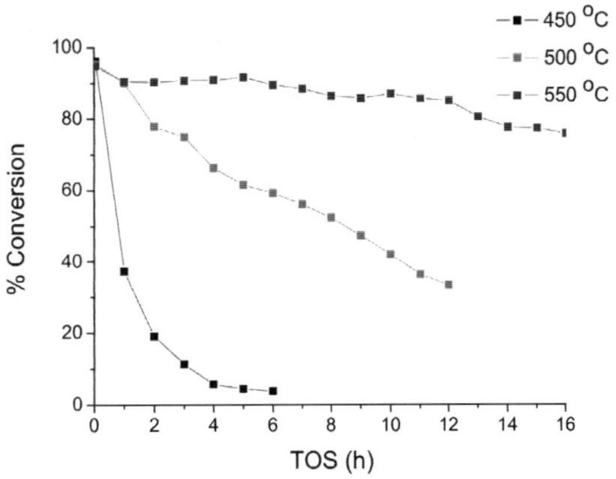

Fig. 5 Deactivation profile of benzaldehyde conversion with various reaction temperature over 3Ga/SiO_2.

(W/F = 250 g.h/mol, reaction temperature 450-550 °C, 100 % H_2 concentration, total flow 30 mL/min)

In addition, it was found that (Fig. 5), a rapidly deactivation can be obtained at low temperature (450 °C). It is consistent with the proposed reaction pathway that, at low temperature, the coupling can be promoted but the decarbonylation/hydrogenolysis activity are readily reduced. This leads to the formation of high molecular weight oxygenates and consequently a rapid catalyst deactivation (< 2 h). As the temperature was increased, such high molecular weight oxygenates can be decarbonylated/hydrogenolysed to lighter aromatic products (benzene and toluene). According, the catalyst stability can be improved.

Effect of H_2 concentration

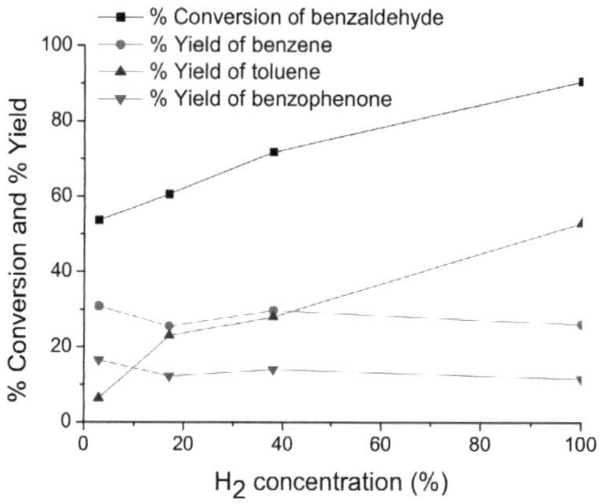

Fig. 6 The effect of H_2 concentration on benzaldehyde conversion over $3Ga/SiO_2$

(W/F = 250 g.h/mol, reaction temperature 550 °C, 3-100 % H_2 concentration, total flow 30 mL/min, the data was shown at 1 hour on stream)

From Fig. 6, the high benzaldehyde conversion was increased with H_2 concentration. This is evidenced that H_2 can play important role on hydrogenolysis of the diketone and also the benzophenone. As H_2 is increased, the high molecular weight oxygenates can be removed leaving more available sites for benzaldehyde activation. Moreover, it is interesting to note that at relatively low H_2 concentration (< 17 %), the selectivity of toluene/benzene ratio < 1. This is clearly explained that the hydrogenolysis activity was diminished due to the low H_2 concentration. Only decarbonylation can be promoted but in much lower extent, as compared to that in the presence of H_2.

Effect of Ga content

From Fig. 7, the conversion of benzaldehyde is increased with Ga content. It is worthy noting that the selectivity of benzene and toluene is similar for 1 and 3 wt% Ga loading. However, it is clearly seen that the toluene/benzene ratio is more than 1 in the case of 6 wt% Ga loading ($6Ga/SiO_2$). This is because the high Ga content can increasingly promote the hydrogenolysis activity leading to the high yield of toluene, as shown.

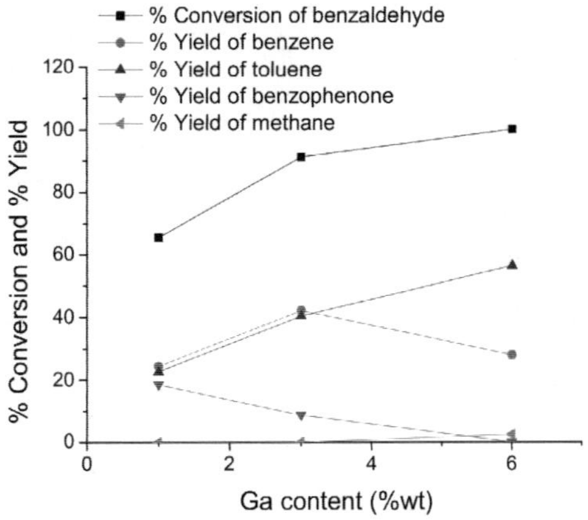

Fig. 7 Effect of Ga content of benzaldehyde conversion.

(Ga loading = 1-6 wt%, W/F = 200 g.h/mol, reaction temperature 550 °C, 100 % H_2 concentration, total flow 30 mL/min, the data was shown at initial conversion)

Effect of Ga cluster

From Table 1, it was found that there is no benzaldehyde conversion over *b*-3Ga/SiO_2. This is presumably because the bulk Ga_2O_3 cannot be readily reduced at reaction temperature. Consequently, the hydrogenation activity was diminished. This suggestion is consistent with the H_2-TPR (Fig. 8). It was found the *b*-3Ga/SiO_2 can be reduced at higher temperature (~730 °C), as compared to 3Ga/SiO_2 which prepared by $Ga(NO_3)_3$ impregnation (~630 °C).

Table 1. Effect of Ga cluster on benzaldehyde conversion over 3Ga/SiO_2.

Catalyst	% of Ga content	% Conversion	% Yield*		
			Benzene	Toluene	Benzophenone
3Ga/SiO_2	3 wt%	91.31	42.18	40.45	8.68
b-Ga/SiO_2	3 wt%	no conversion	-	-	-
reduced-*b*-Ga/SiO_2	3 wt %	29.04	7.59	10.33	11.41

(W/F = 200 g.h/mol, Reaction temperature 550 °C, 100 % H_2 concentration, Total flow 30 mL/min, Ga content = 3 wt%, the data was shown at initial time on stream)

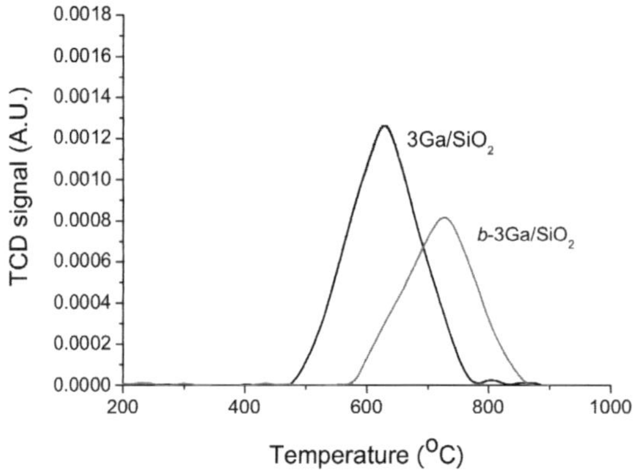

Fig. 8 TPR profile of the $3Ga/SiO_2$ and *b*-$3Ga/SiO_2$ catalyst

It was believed that the higher reduction temperature of *b*-$3Ga/SiO_2$ is due to the larger cluster size of Ga_2O_3 particle, as compared to the Ga species prepared by impregnation. The lower reduction temperature peak obtained from $3Ga/SiO_2$ indicates the well-dispersed Ga_2O_3 particles in this sample.

When *b*-$3Ga/SiO_2$ was then reduced at 700 °C for 2 hours under 100 % H_2, and then cooled to 550 °C under 100% H_2. The activity of this activated catalyst is dramatically increased (Table 1). It is suggested that cluster size of Ga species play role on hydrogenation/hydrogenolysis activity.

Conclusions

The benzaldehyde conversion over Ga supported silica at atmospheric pressure and 450-550 °C produced benzene, toluene, and benzophenone. Stability and selectivity of the products depends largely on the hydrogenolysis and decarbonylation activity of the catalyst. The deactivation of the decarbonylation activity leads to high toluene/benzene yield. While the hydrogenolysis activity can particularly enhance the toluene/benzene ratio. Inadequate H_2 concentration and temperature leads to a rapid catalyst deactivation due to the formation of high molecular weight oxygenates. Increase in Ga content result in a higher hydrogenolysis activity and stability.

Acknowledgements

The authors are grateful for a financial support from the Thailand Research Fund through the Royal Golden Jubilee Ph.D. Program (Grant No. PHD/0213/2548).

References

1. Saadi A., Merabti R., Rassoul Z. and Bettahar M.M. (2006), "Benzaldehyde Hydrogenation over Supported Nickel Catalysts," Journal of Molecular Catalysis A: Chemical, 253, pp. 79-85.

2. Gerardo F. Santori, Mónica L. Casella and Osmar A. Ferretti, (2002), "Hydrogenation of Carbonyl Compounds Using Tin-Modified Platinum-Based Catalysts Prepared via Surface Organometallic Chemistry on Metals (SOMC/M)," Journal of Molecular Catalysis A: Chemical, 186, pp. 223-239.
3. Dana Procha´zkova´, Petr Za´mostny´, Martina Bejblova´, Libor Cˇerveny´ and Jirˇı´ Cˇejka, (2007), "Hydrodeoxygenation of Aldehydes Catalyzed by Supported Palladium Catalysts," Applied Catalysis A: General, 332, pp. 56-64.
4. Saadi A., Rassoul Z. and Bettahar M.M., (2000), "Gas Phase Hydrogenation of Benzaldehyde over Supported Copper Catalysts," Journal of Molecular Catalysis A: Chemical, 164, pp. 205-216.
5. Duncan T. Burns, Kanita Tungkananurak and Damrongsak Jadsadapattarakul, (2006), "Semi-Micro Preparation and Characterization of Bonded Phase ODS-Silica Prepared from Rice Husk Silica," Microchim Acta, 154, pp. 81-85.
6. Kanita Tungkananurak, Sirinlux Kerdsiri, Damrongsak Jadsadapattarakul and Duncan Thorburn Burns, (2007), "Semi-Micro Preparation and Characterization of Mesoporous Silica Microspheres from Rice Husk Sodium Silicate Using a Non-Ionic Surfactant as a Template: Application in Normal Phase HPLC Columns," Microchim Acta, 159, pp. 217-222.
7. Judith M. Percino, V´ıctor M. Chapela, Omar Urz´ua and H´ector Toribio, (2007), "Decarbonylation of Two *alpha*-Diketones: 1,2-Bis(6-Methylpyridin-2-yl)Ethane-1,2-Dione and 1-(Pyridin-2-yl)-2-(6-Methylpyridin-2-yl)Ethane-1,2-Dione," International Journal of Mass Spectrometry, 262, pp. 122-127.

634d Steam Reforming of Different Fractions of Biomass Pyrolysis Liquids

Pedro J. Ortiz-Toral[1], Justinus A. Satrio[2], Robert C. Brown[2], and Brent H. Shanks[1]. (1) Department of Chemical and Biological Engineering, Iowa State University, Ames, IA 50011, (2) Center for Sustainable Environmental Technologies, Iowa State University, Ames, IA 50011

Hydrogen can be obtained from biomass pyrolysis liquids by catalytic steam reforming. Catalyst deactivation by coking and the formation of carbon deposits are the major known limitations although the specific causes are unidentified. It is proposed that the known limitations could be reduced by selectively reforming specific fractions of the bio-oil. The hydrophobic fraction composed of phenolic and furan derived compounds can be separated by water extraction as a high-value product. In the current work, the particular interest is on efficient hydrogen production from the residual aqueous fractions. There are certain challenges introduced due to the complexity of these mixtures. An improved understanding of the underlying steam reforming characteristics of pyrolysis liquids compounds will provide the opportunity to predict hydrogen yields and to potentially identify undesired compounds. Different fractions were being studied to determine which ones can be used more efficiently to produce hydrogen with existing steam reforming technologies. A fractionating condenser system was used to collect different fractions of fast pyrolysis liquids. The hydrophilic components from the different fractions were analyzed and steam reformed using a nickel-based commercial catalyst loaded fixed bed reactor. The results from these studies are used to correlate the steam reforming efficiency to the characteristics of the fractions. The performance of a combined catalyst-sorbent will also be discussed in comparison to hydrogen yields obtained from bio-oil with the commercial catalyst.

634e Millisecond Autothermal Catalytic Partial Oxidation of Glycerol to Syngas and Chemicals

David Rennard and Lanny D. Schmidt. Chemical Engineering and Material Science, University of Minnesota, Amundson Hall, 421 Washington Ave, Minneapolis, MN 55455

Partial oxidation of glycerol occurred over noble metal catalysts in millisecond contact times. Glycerol and air were fed into a quartz reactor at room temperature through an entrained flow nebulizer. This liquid-gas mixture passed over a Rh-Ce catalyst, reacting at temperatures of 500 C to 1000 C with near 100% conversion and high selectivity to synthesis gas. Water was added to the liquid glycerol at up to a 5:1 molar ratio. Neat glycerol was also converted into chemicals over a Pt catalyst: acrolein, hydroxyacetone, acetaldehyde, and ethylene were observed.

634f A Comprehensive Look at the Acidic Hydrolysis of Loblolly Pine Sawdust

Teresita Marzialetti, Mariefel Valenzuela-Olarte, Carsten Sievers, Travis Hoskins, Pradeep K. Agrawal, and Christopher W. Jones. School of Chemical & Biomolecular Engineering, Georgia Institute of Technology, 311 Ferst Drive NW, Atlanta, GA 30332-0100

A comprehensive study of the acid hydrolysis of Loblolly pine, using different hydrolysis conditions is presented. The effect of the type of acid, pH, reaction temperature, and reaction time on hydrolysis products such as monosaccharides and the subsequent degradation products, HMF and furfural, is reported using a stirred batch reactor. Trifluoroacetic acid is found to yield the highest amount of overall soluble monosaccharides mainly from hemicelluloses. The mineral acids (HCl, H2SO4, HNO3, H3PO4) gave a slightly lower yield of monosaccharides from hydrolyzed hemicellulose because monosaccharides are faster degraded. At 200°C, substantial amount of cellulose is hydrolyzed by mineral acids. Larger amounts of degradation products are also noted at higher temperatures. TFA was found to be the most "gentle" acid, leading to limited monosaccharide degradation among the acids used. The presence of soluble oligosaccharides in solution after hydrolysis was confirmed by applying a secondary acid hydrolysis to the solid-free liquid hydrolysate.

634g Biogas: Sludge to Biogas

Chikezie N. Nwaoha, Petroleum Engineering, Federal University of Technology Owerri, Nigeria, 10a lake nwaebere aladinma, Owerri, 234, Nigeria

Energy conservation has become a major case in all energy related industry, because of the rising demand and shrinking reserves of energy.

Sludge, a precipitated solid matter gotten from the treatment of waste water can be further treated to generate a gas (sludge gas). This is a type of biogas which is produced from the anaerobic digestion of the organic substance sludge, which can serve as a superior alternative to composting such type of biomass. The biogas consists of about 60-70% of methane and 30-40% of carbon dioxide. This composition makes it highly suitable for combustion in gas engines to generate thermal energy for heating the sludge, offsetting other heating requirements, and generating electricity.

The treated water from the plant can also serve as a make-up water for the boiler to generate steam for SAGD (Steam Assisted Gravity Drainage) and non-SAGD applications.

Besides its economic advantages, it also has a lesser impact on the environment and helps to conserve valuable resources.

685a Product Distribution from Fast Pyrolysis of Polysaccharides

Pushkaraj R. Patwardhan[1], Justinus A. Satrio[2], Brent H. Shanks[1], and Robert C. Brown[2]. (1) Department of Chemical and Biological Engineering, Iowa State University, Ames, IA 50011, (2) Center for Sustainable Environmental Technologies, Iowa State University, Ames, IA 50011

Polysaccharides constitute the largest fraction of biomass. With the upsurge of interest in utilization of biomass as a renewable resource for fuels and chemicals via the fast pyrolysis route, studying carbohydrate pyrolysis is of key importance. Several researchers have presented pyrolysis models that can predict the yields of overall char, bio-oil and gaseous fractions; however, models have yet to be developed that can predict the more detailed composition of the bio-oil. Furthermore, little work has been done to distinguish between the products of primary and secondary reactions. In the present study, several mono-, di- and polysaccharides, varying in the position and orientation of glycosidic linkages, were pyrolyzed under conditions that minimized secondary reactions. The product distribution from the pyrolysis of these polysaccharides tells us that α or β orientation of the glycosidic linkage has negligible effect on the product distribution, however the position of linkage has a significant impact on product yields.

685b Catalytic Gasification of Biomass in Supercritical Water

Fernando L. P. Resende and Phillip E. Savage. Chemical Engineering Department, University of Michigan, H.H. Dow Building, 2300, Hayward St., Ann Arbor, MI 48109-2136

The use of renewable sources of energy such as biomass is an alternative that has the potential to reduce our reliance on imported oil while addressing environmental concerns. Among the technologies that convert biomass into fuels, Supercritical Water Gasification (SCWG) has been suggested to process wet feedstocks because of the ability of water to dissolve organic components of plant materials at supercritical conditions, promoting formation of gases such as H_2 and CH_4, and limiting the amount of tar and char formed as residues.

Metal catalysts can improve SCWG efficiency in a variety of ways. Some catalysts promote the cleavage of C-C bonds, which is crucial to achieve higher gasification yields. Others promote reactions in the gas-phase changing product composition towards H_2 production, such as water-gas shift. This research project aims to determine which metal catalysts are more effective to conduct SCWG, and the conditions where catalytic effects are maximized.

We gasified cellulose and lignin as model compounds for biomass in supercritical water, and avoided catalytic effects from the reactor walls by using quartz reactors. We ran experiments in the absence of catalysts and in the presence of nickel, iron or copper wires to evaluate their effects on product yields. In order to determine how experimental conditions affected catalysis, we varied temperature from 400 to 725 °C, biomass loading from 5.0 to 33.3 wt %, and water density from 0.08 g/ml to 0.18 g/ml.

The results indicate that high temperatures, low loadings and high water densities promoted catalytic activity. Nickel increases yields of the gas products (H_2 in particular) at all conditions studied, while iron and copper increase yields only at specific conditions. When CH_4 and energetic yields are also taken into account, iron is more effective at high temperatures and copper at low biomass loadings.

685c Investigation of High Temperature Gasification of Biomass In An Aerosol Flow Reactor

R. Bryan Woodruff[1], Christopher Perkins[1], and Alan W. Weimer[2]. (1) Department of Chemical and Biological Engineering, University of Colorado at Boulder, 1111 Engineering Drive, Boulder, CO 80309-0424, (2) Department of Chemical and Biological Engineering, University of Colorado, 1111 Engineering Drive, Boulder, CO 80309-0424

Steam gasification of biomass to produce synthesis gas (CO, CO2, H2) was performed at high temperature (1273-1623 K) in an aerosol flow reactor. High temperatures and rapid heating lead to fast reaction kinetics, high conversion efficiencies, and very low tar production. The formation of syngas from biomass is a critical reaction in a promising method to produce liquid fuels from a renewable energy source. Current methods of gasification burn a portion of the biomass to provide the heat of reaction which decreases the quantity and quality of the syngas. Concentrated solar thermal energy offers a renewable alternative that can greatly increase the syngas quality and yield. Fundamental understanding of reaction parameters on conversion efficiency is needed to optimize the solar reactor design. An electrically heated aerosol reactor was constructed to investigate the effects of temperature, particle size, carrier gas flow rate, and steam addition on the carbon conversion efficiency in a controlled environment. Conversion efficiencies ranging from 6% to 88% were observed using corn stover at residence times on the order of 0.5 seconds with a particle size ranging from 150-2000 μ m.

685d Biomass Gasification: Catalytic Tar Removal Using Ni-Zeolite Catalysts

Prashanth R. Buchireddy, Mississippi State University, P.O.Box 9595, 330 Swalm Chemical Engineering Building, Starkville, MS 39762 and R. Mark Bricka, Chemical Engineering, Mississippi State University, Box 9595, Mississippi State, MS 39762.

Biomass is considered as a potential feedstock for sustainable energy production that can be converted to chemicals, fuels, electricity etc., either by thermochemical or biological processes. Gasification is one of the thermochemical methods of converting biomass to synthetic gas (CO and H2), which could potentially be used to produce different chemicals such as acetic acid and ethanol via chemical or biological processes and/or generate electricity by using engines, turbines, fuel cells etc. However, the synthetic gas produced may contain different organic (Tars) and inorganic impurities, which have the potential to damage the process equipment used in different conversion processes. Hence, the synthetic gas has to be cleaned to remove these impurities, and the extent to which the gas has to be cleaned is dictated by the end use application.

Research is being carried out to evaluate different nickel supported Zeolite-Y catalysts for tar destruction. Nickel supported Y-Zeolites with different SiO2/Al2O3 ratios (5.2, 30, 80) were tested to evaluate the effect of catalyst acidity on tar removal in the temperature range of 550-750 OC. Also, nickel supported on Silica/Alumina catalyst was prepared and tested. Catalyst characterization was performed using XRD, BET, and SEM on the catalysts evaluated. Tests were carried out on a laboratory scale flow reactor with naphthalene as a tar model compound. The results of these experiments showed that Ni on Zeolite-Y had better naphthalene removal efficiencies compared with Silica/Alumina catalyst, which could be attributed to the higher surface area and acidity of the zeolite. Also, the acidity of the zeolite support had a significant effect of naphthalene removal efficiencies. Removal efficiencies decreased with an increase in the acidity of the catalyst. This could be due to the formation of coke and NiAl2O4, which was confirmed by the TGA and XRD analysis. Tests were also carried out to evaluate the effect of varying nickel loading on zeolite, naphthalene loading, steam to carbon ratio, and gas hourly space velocity, the results of which will be presented.

685e Low-Temperature Gasification of Cellulosic Biomass

Sameer Parvathikar and Johannes W. Schwank. Chemical Engineering, Transportation Energy Center, University of Michigan, 2300 Hayward, 3014 H.H. Dow Building, Ann Arbor, MI 48109-2136

Small modular biopower systems offer a low footprint and cost effective solution to existing problems like global deforestation and increased food prices, in large scale biomass usage. A broad range of technologies exist for the conversion of biomass into useful products. Gasification and pyrolysis has traditionally been carried out at very high temperatures and the catalysts used have been targetted at tar reduction and syn-gas quality product formation. Previous research has also focussed on evaluating products on the basis of total gas formation and total oil formation. Significant gaps exist in the understanding of low temperature gasification and product analysis.

Thermogravimetric analysis of biomass decomposition has been carried out in different atmospheres and in the presence of different primary catalysts. The main focus of the research has been on the identification and quantification of the products of biomass gasification and pyrolysis according to their functional groups as a function of the temperature and time of evolution. This would enable the identification of volatiles evolved at low temperatures and and investigate the feasibility of using these volatiles to make useful products. The gases have been analysed by gas chromatography and fourier transform IR spectroscopy and the liquid products have been analysed by a GC/MS. The effect of different primary and secondary catalysts in altering product composition and yields are also discussed.

685f Mechanism of Thermal Decomposition of Furan

AnGayle K. Vasiliou[1], Hans-Heinrich Carstensen[2], Mark R. Nimlos[3], John W. Daily[4], and G. Barney Ellison[1]. (1) Chemistry, University of Colorado at Boulder, Boulder, CO 80309, (2) Chemical Engineering Dept, Colorado School of Mines, 1613 Illinois St., Golden, CO 80401, (3) National Renewable Energy Laboratory, 1617 Cole Boulevard, Golden, CO 80401, (4) Department of Mechanical Engineering, University of Colorado at Boulder, Center for Combustion and Environmental Research, Boulder, CO 80309

Both furan (C_4H_4O) and furfural (C_4H_3O-CHO) are important products in biomass pyrolysis. To understand the thermal cracking of furan, we have studied its thermal decomposition by entraining a few Torr in 1–2 atmospheres in a high temperature supersonic nozzle. Furan samples are heated to temperatures of 1000 – 1400 K for 30–50 μsec. The decomposition products are identified by photoionization mass spectroscopy and infrared spectroscopy.

In the hyperthermal nozzle, the organic species such as furan are decomposed while entrained seeded in an inert gas (roughly 1 – 2 atm) through a resistively heated silicon carbide tube (1 mm ID, about 2 cm long, temperature up to 1800 K) into a high vacuum chamber. The injection of the gas pulses into the tube is supersonic, which is achieved by an increase in cross-sectional area upon going from the orifice in the pulsed valve (0.1 - 0.4 mm diameter) into the SiC tube. The near-sonic gas flow velocity through the tube allows a short contact time (roughly 30 μsec), which is essential to avoid radical-radical reactions. The jet of organic fragments and the He carrier gas expands into a vacuum at 10^{-5} Torr. Subsequently, the beam of radicals impinges onto a 20 K CsI window forming a matrix for IR detection. Photoionization time-of-flight (TOF) mass spectroscopy is used to detect the species coming out of the nozzle and optimize the thermal dissociation conditions such as nozzle temperature, and length for maximum radical production. The TOF mass spectrometer uses the 9th harmonic of a YAG laser at 118.2 nm (or 10.487 eV) to photoionize the target species.

The thermal fragmentation of furan has been studied previously. In an early study, furan was passed through a heatable molecular flow reactor operating at low pressures of about 1 mTorr. Product analysis was continuously performed by an on-line quadrupole mass spectrometer with electron impact ionization. Furan was observed to decompose between 1050 and 1270 K by ring breakdown unimolecular reactions. Loss of carbon monoxide was the exclusive process. It was found that $k_o(1)$ is $10^{15.6}$ exp(-74 kcal mol^{-1}/RT).

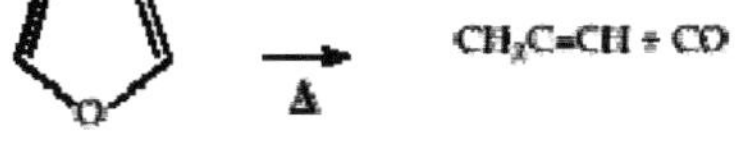

(1)

Instead of a flow tube, the thermal decomposition of furan was studied behind reflected shocks in a single shock tube, over the temperature range 1050-1460 K. Shocked samples were taken from the end block of the driven section and were analyzed on a Packard 800 series gas chromatograph using a flame ionization detector. Methylacetylene and carbon monoxide were the major reaction products of the furan decomposition reaction: CH_3C¼CH + CO. The rate constant was measured to be $10^{15.25\pm0.5}$ exp[-(77.5 ± 2.5) kcal mol^{-1}/RT] s^{-1}. In addition to (1), a second initiation reaction produces HCCH and CH_2CO with a rate constant of $10^{14.7\pm0.5}$ exp [-(77.5 ± 2.5) kcal mol^{-1}/RT] s^{-1}.

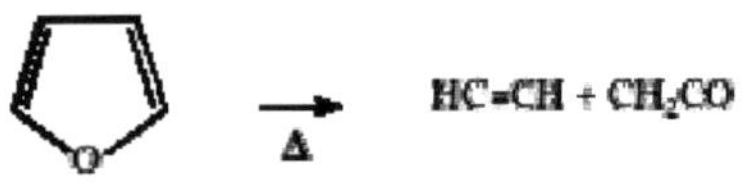

(2)

Additional reaction products which appear in the pyrolysis were identified by their GC traces to be: CH_2=C=CH_2, C_4H_6, CH_2=CH_2, CH_4, HCC-CCH, and C_6H_6.

In a subsequent shock tube study, the vibrational relaxation, incubation times, and unimolecular dissociation of C_4H_4O was investigated over the extended temperature range 500-3000 K in 2-5% furan-krypton mixtures, 2% furan-neon mixtures, and in pure furan at pressures of 600 Torr. The experiments were performed in shock waves using laser-schlieren densitometry and time-of-flight mass spectrometry. Unimolecular dissociation is observed in mass spectrometry experiments between 1300 and 1700 K in a pressure range of 175-250 Torr as well as laser-schlieren experiments between 1700 and 3000 K for pressures between 100 and 600 Torr. The mass spectrometry experiments show that under the given conditions two molecular dissociation channels leading to (1) C_3H_4 + CO or to (2) HC¼CH + CH_2CO are dominant. The branching ratio between these channels has been determined between 1300 and 1700 K. At low temperatures, the molecular channel leading to C_3H_4 and CO is preferred, but a channel switching was observed around 1700 K.

When we decompose furan in a hyperthermal nozzle, no fragmentation is observed until nozzle temperatures of 1000 K are achieved. Residence times in the supersonic nozzle are 30–50 μsec.

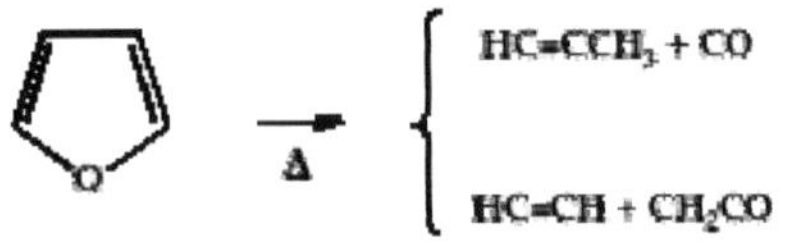

(3)

The reaction products are identified by features in the PIMS at m/z 40 and 42. In separate experiments, infrared signatures of $HCCCH_3$, CO, HCCH, and CH_2CO have been detected when the reaction products are trapped in a cryogenic matrix. As the nozzle temperature is raised to 1400 K, the PIMS spectrum shows peaks at m/z 39 (the propargyl radical, $HCCCH_2$) and m/z 41 (ketenyl radical, HCCO, or allyl radical, CH_2CHCH_2).

If the concentration of furan in the fuel mixture is raised by a factor of 10, the PIMS spectrum reveals features at m/z 78 (benzene), m/z 94 (probably phenol, C_6H_5OH), and m/z 104 (perhaps styrene, $C_6H_5CHCH_2$). In earlier work, at high nozzle pressures we have observed dimerization of propargyl radicals to generate benzene ($HCCCH_2$ + $HCCCH_2$ ¨ C_6H_6). Consequently we believe that radical/ radical processes are generating these higher molecular weight products.

Ab initio electronic structure calculations have been carried out and these CBS-QB3 applications reveal that the a-C-H bond energy in furan is greater than benzene, DH_{298}(a-C_4H_3O-H = 120 ± 2 kcal mol^{-1}). Because the C-H bonds in furan are so strong, the initial step in the cracking of furan is likely rearrangement to a carbene which subsequently fragments to products. The mechanisms for furan cracking are certain to be complicated.

AUTHOR INDEX

Abunasser, Nadia J. 18, 59, 123
Agrawal, Pradeep K. 192
Akkapeddi, Srikanth 113
Alfstad, Thomas 31
Alley, Earl 22, 75
Amen, Q. V. 149
Antal, Steven P. 60
Ausavasukhi, Artit 181
Bain, Richard L. 31
Balakrishnan, Nianthrini 144
Beaird, Amy M. 141
Benoit, Jeremiah 86
Bhethanabotla, Venkat 144
Bommareddy, Susilpa 88
Boovaragavan, Vijayasekaran 127
Bricka, R. Mark 197
Brown, Robert C. 190, 194
Buchireddy, Prashanth R. 197
Cao, Dapeng 139
Carstensen, Hans-Heinrich 199
Castaldi, Marco J. 89
Chandrasekaran, Rajeswari 129
Chappell, Joeseph 19
Chiang, Benjamin 9
Choudhury, Pabitra 144
Crossley, Steven P. 157
Curtis, Wayne 19
Daggolu, Prashant Reuben 158
Daily, John W. 199
Danuthai, Tanate 159
Dathara, Gopi Krishna Phani 140
Davis, Thomas A. 141
De Guzman, Rhet Joseph 107, 108
Deng, Shuguang 21, 156
Denis, Jonathan 129
Doty, David 78
Du, Chenyu 23
Duke, Steve R. 113
Dumesic, James A. 180
Eden, Mario R. 88
Ekundayo, Babatunde 142
Ellison, G. Barney 199
Estabrook, Benjamin D. 167
Farag, Ihab H. 1, 9
Ford, Andrew T. 22
French, Todd 75
French, William T. 22
Fuller, Thomas F. 129
Gabauer, Steve 19
Gomez, Jorge Mario 91
Gribik, Anastasia M. 60
Guillen, Donna Post 60
Hall, Jaclyn 75
Hashem, Danah 114
Hernandez, Rafael 22, 75
Holmes, William 22, 75
Hoskins, Travis 192
Huang, Simon 114
Immer, Jeremy G. 167
Jones, Christopher W. 192
Joseph, Babu 144
Kelly, M. Jason 167
Khatri, Waqas 19
Kim, Manhoe 123
Kitiyanan, Boonyarach 109
Kunkes, Edward L. 180
Kwon, Kyung C. 168
Kyriacopulos, Paul C. 1
Lamb, H. Henry 167
Lan, Jianhui 139
Leach, Ryan 1
Lee, Audrey 31

AUTHOR INDEX

Lo, Chi 90
Lobban, L. L. 149, 159
Mainardi, Daniela S. 140
Mallinson, R. G. 149, 159
Marolla, Theordore 168
Marrero-Alfonso, Eyma Y. 141
Marzialetti, Teresita 192
Mashburn, Michael 168
Matthews, Michael A. 141
Mayfield, Howard 168
Meeyoo, Vissanu 87, 109
Miao, Shaojun 148
Mondal, Kanchan 86
Ng, K. Y. Simon 18, 59, 107, 108, 123
Nimlos, Mark R. 199
Nwaoha, Chikezie N. 193
Orduz, Federico Velosa 91
Ortiz-Toral, Pedro J. 190
Osborne-Lee, Irvin W. 142
Osuwan, Somchai 159
Ouellette, Jason 1
Parvathikar, Sameer 198
Patil, Prafulla 21, 156
Patwardhan, Pushkaraj R. 194
Payne, Robert U. 124
Pearson, Joseph 114
Peralta, Maria A. 159
Perez, Mario Enrique Danton Garcia 59
Perkins, Christopher 196
Rangsunvigit, Pramoch 87, 109
Rennard, David 191
Resasco, Daniel E. 149, 157, 159, 181
Resende, Fernando L. P. 195
Rirksomboon, Thirasak 109
Ruiz, Maria Pilar 157
Salley, Steven O. 18, 59, 107, 108, 123
Sammons Jr., Norman E. 88
Sastri, Bhima 31
Satrio, Justinus A. 190, 194
Savage, Phillip E. 195
Sawyer, Bryan 138
Schindler, Anton 113
Schmidt, Lanny D. 191
Schroer, Josef G. 77
Schwank, Johannes W. 198
Shanks, Brent H. 148, 190, 194
Shelley, Jonathan K. 60
Shevgoor, Siddarth 78
Sievers, Carsten 192
Simonetti, Dante A. 180
Skinner, Michael 129
Sooknoi, T. 149, 159, 181
Stafford, Don 113
Subramanian, Venkat 127
Suppes, Galen J. 138, 179
Swart, Dustin 113
Taboada-Serrano, Patricia 58
Tang, Haiying 59, 107, 108, 123
Tatarchuk, Bruce J. 124
Tereck, Charles D. 76
Theodoropoulos, Constantinos 23
Thomas, J. Mathew 75
Tsouris, Costas 58
Ung-Jinda, Charothon 109
Valenzuela-Olarte, Mariefel 192
Vasiliou, Angayle K. 199
Vlysidis, Anestis 23
Walker, Sean 142
Wang, Wenchuan 139
Webb, Colin 23
Webster, Thaddaeus Andrew 9
Weimer, Alan W. 196

AUTHOR INDEX

West, Ryan M. **180**

White, Mark G. **158**

Wilson, John **18, 59, 123**

Wiltowski, Tomasz **86**

Woodruff, R. Bryan **196**

Yan, Shuli **123**

Yan, Wei **179**

Yegulalp, Tuncel **89**

Yuan, Wei **88**

Zhang, Liang **157**

Zhou, Yue **89**

Zhu, Wenhua H. **124**

Zhu, Ying **124**

American Institute of Chemical Engineers
3 Park Avenue
New York, New York 10016-5991

CURRAN ASSOCIATES INC
proceedings.com